ART DECO ALUMINUM

Kensington

Paula Ockner
& Leslie Piña

Schiffer Publishing Ltd

4880 Lower Valley Road, Atglen, PA 19310 USA

Acknowledgements

We would like to extend our thanks to many people who shared information about Kensington and/or led us to sources, especially Bonita Campbell, Steve Doell, Chris Kennedy, Ralph and Terry Kovel, Robert G. Long, Ann Madarasz, Paul Peoples, Paul Roberts, Gregory Smith, and Barbara Stewart, and to organizations, notably the Historical Society of Western Pennsylvania and, of course, Alcoa for generously giving access to their archives and permission to use vintage photographs. We are especially grateful to Barbara Stewart at Alcoa for her time and friendly assistance, to the staff in Government Documents at the Cleveland Public Library who helped make our search for patents almost bearable, and to the librarians of Special Collections of Bird Library at Syracuse University for graciously making the voluminous Lurelle Guild Archives available to us.

Thanks to Paula's husband Steve, for caring enough about his home town and about good art to collect Kensington Ware long before it became popular again, to Leslie's husband Ramón for his help with the photography, with paging through volumes of *The Alcoa News*, and for getting dizzy advancing the library microfilm reader; and thanks again to Doug and the staff at Schiffer Publishing for their continued support and friendship.

Dedicated to the memory of Lurelle Guild and to the people at Alcoa who left us Art Deco aluminum

Ockner, Paula.
    Art deco aluminum : Kensington / Paula Ockner & Leslie Piña .
        p.   cm.
    Includes bibliographical references.
    ISBN 0-7643-0366-X
    1. Aluminum giftware -- Pennsylvania -- New Kensington -- Catalogs. 2. Decoration and ornament -- Pennsylvania -- New Kensington -- Art deco -- Catalogs. I. Piña, Leslie A., 1947-   . II. Title.
NK7700.026   1997
9.5'7'075--dc21   97-21753
                        CIP

Published by Schiffer Publishing Ltd.
4880 Lower Valley Road
Atglen, PA 19310
Phone: (610) 593-1777; Fax: (610) 593-2002
E-mail: schifferbk@aol.com
Please write for a free catalog.
This book may be purchased from the publisher.
Please include $3.95 for shipping.
Try your bookstore first.

We are interested in hearing from authors
with book ideas on related subjects.

Contents

Introduction

Art Deco is a convenient label for a complex range of styles that evolved from an earlier Parisian phenomenon called Art Nouveau. An elitist, eccentric attempt at a "new art" (Art Nouveau) in Europe was transformed into an elitist, eccentric "modern art" (Art Moderne) at the time of the First World War. Although Art Moderne had its spirit in the present modernizing world, its body had at least one foot in the past. Sumptuous objects were handcrafted by techniques reminiscent of French ébénistes in the courts of the Louis. Design had the distinct flavor of classicism along with its later interpretive revivals. Exotic tropical woods, ivory, and metals embellished the furnishings of the fashionable wealthy class. French Art Moderne, as shown in Paris at *L'Exposition des Arts Decoratifs et Industriels Modernes* in 1925, was not yet suited as an international style.

The United States was noticeably absent at the milestone event that introduced modern decorative and industrial arts to the design-conscious world. When Secretary of Commerce Herbert Hoover received the French invitation to show American modern art, he declined. Had he missed something? To find an answer, Hoover appointed a group of delegates from manufacturing, retail, and decorating fields to visit the exposition and report back on what America was missing. Their return home from Paris late in 1925 marked a significant beginning to a transformed, commercialized, Americanized Art Moderne, or as it is now called, Art Deco.

Art Deco entered America at the port of New York in the display windows of major department stores. First fashion—clothing, accessories, perfume bottles, and furnishings used as props—caught the eye and the imagination of the American consumer. Originally intended to be relatively inaccessible and absolutely costly, items were now desired for their appearance rather than for materials and crafting techniques. The new look was no more costly than the tired historic and bland generic "styles" it soon replaced. Art Deco graphic design quickly appeared in periodicals and posters, and it was equally suited to the industrial design of useful objects. Infatuated by mass-production, speed, metallic sparkle, and the new age that they represented, American designers took the French look and ran with it. From architecture to plastic jewelry, American Art Deco had its spirit in the future and its body in a rapidly transforming present.

By 1925 France and other participating European nations had grown tired of the style with design cliches of leaping gazelles, muscular nudes, and kinky sexual innuendo. In the late 1920s and 1930s the United States carried on with its own cliches and promises of a brave new world—rounded corners and three parallel lines signifying streamlining, a stepped or tiered pattern resembling a skyline of skyscrapers, and an Americanized athletic female dancing nude. The geometry of Cubism replaced the representational motifs of tradition. With added ingredients from ancient Egypt, pre-Columbian Mexico, some less-explainable European peasant art, and the austere simplicity of the German Bauhaus, the American Art Deco stew was ready for mass consumption. This meant, of course, that mass, or at least factory, production and marketing were needed. America was the right place, and the aluminum industry was as suited as any to participate in the version of the Art Deco style variously referred to as "machine art" or "machine-age modern" or "Streamlined Modern" or simply "American Art Deco."

In a catalog for an exhibition called "Machine Art," the author quotes from Plato's *Philebus* on the beauty of shapes: "...straight lines and circles, and shapes, plane or solid, made from them by lathe, ruler and square. These are not, like other things, beautiful relatively, but always and absolutely." And from St. Thomas Aquinas's requirements for beauty in *Summa Theologiae*, "...integrity or perfection: those things which are broken are bad for this very reason...due proportion or harmony...clarity: whence those things which have a shiny color are called beautiful."

Alfred Barr Jr. applies these ideas to machine art in his Foreword: "The beauty of machine art is in part the abstract beauty of straight lines and circles." He asserts that modern 1930s materials and precision instruments have enabled the machine-made object to approach or participate in Plato's ideal of pure shapes. Beauty of surface is another quality of good machine art. "Machine art, devoid as it should be of surface ornament, must depend upon the sensuous beauty of porcelain, enamel, celluloid, glass of all colors, copper, aluminum, brass and steel." The designer simply chooses the most appropriate form based on both functional and aesthetic criteria. "He does not embellish or elaborate, but refines, simplifies and perfects."

What developed as a result of this modern merger between art and industry (and philosophy) was a new field of industrial design, especially in America in the 1930s. The initial goal of industrial design was to transform ordinary things into extraordinary objects of desire by making them very visually attractive and affordable to a large audience. It represented the antithesis of elitist handicraft, because industrial design used mass production, meaning machinery, for everything, including art. Its streamlined style, however, was based not only on aerodynamics but on French Art Deco.

The irony is twofold: the products were usually unrelated to transportation—why make a toaster with aerodynamic styling?—and Art Deco design originally focused on luxury art and not utility. Yet, in many instances, it worked, and it worked well.

Lurelle Guild was one of the American pioneer professional industrial designers to translate aspects of French Art Deco to American vernacular and methods of mass production. Among his many contributions was a new collection of smooth, shiny aluminum decorative and utilitarian items made by Alcoa in New Kensington, Pennsylvania and called Kensington Ware. Strictly speaking, Kensington Ware and other 'thirties industrial designs are examples of machine age products. But they are not without style, and what becomes apparent when looking at many of Guild's patent drawings and silvery-finished decorative art objects is the French accent on its American modern design vocabulary. Some of his designs are obviously something else—inspired by colonial American and more traditional themes—but those that can be identified with Art Deco style, usually called "classic modern" in their day, are both historically and aesthetically significant.

The Company

Alcoa

Although aluminum is a common element found in the earth's crust, an economical way of separating it from bauxite ore was unknown until the late nineteenth century. In 1886, Charles Martin Hall (1863-1914), a young scientist who had graduated from Oberlin College in Ohio, perfected the electro-chemical reduction process that became the foundation of the aluminum industry. In 1888, he founded the Pittsburgh Reduction Company, better known as the Aluminum Company of America (Alcoa) since 1907. In 1891 a new plant was built nineteen miles north of Pittsburgh, across the Allegheny River in New Kensington. Aluminum items gradually became accepted; Teddy Roosevelt even used an aluminum canteen at San Juan Hill in 1898. Both the Empire State Building and Rockefeller Center in New York have Alcoa aluminum building components. In 1952, Alcoa constructed a spectacular 30-story headquarters in the heart of downtown Pittsburgh, with shiny aluminum exterior panels, doors, windows, elevators, ceilings, light fixtures, and furnishings. Today, a hallway ends with a glass-enclosed display of Kensington Ware.

View of a display cabinet at Alcoa headquarters in Pittsburgh.

Until the 1940s, Alcoa had enjoyed a monopoly on aluminum production in America. The problem had never been in providing the lightweight corrosion-resistant metal, but in convincing other companies to use it. Since Alcoa was primarily a producer of raw aluminum, manufacturing companies needed to be made aware of its potential as a substitute for other materials used to manufacture familiar items. Research and development, finding new markets, and educating the public were as essential for Alcoa's early success as the production of the metal. One of the earliest and most rewarding attempts to promote aluminum was Alcoa's entry into the cooking utensil market.

Wear-Ever

In 1901 a subsidiary company was incorporated as The Aluminum Cooking Utensil Company, Inc., primarily as the sales organization for marketing the aluminum cooking utensil products of the Pittsburgh Reduction Company. The Wear-Ever name was adopted n 1903. Distributed through a direct-to-consumer sales organization, which included college students peddling the wares door-to-door, Wear-Ever Aluminum cooking utensils became the first widely accepted household product fabricated of aluminum. The company name of Wear-Ever Aluminum, Inc. was adopted in 1903. Manufacturing and administration were centered in New Kensington, Pennsylvania.

Wear-Ever Aluminum, Inc. was reorganized January 1, 1966, as an integrated marketing-manufacturing subsidiary of Alcoa. Consumer cookware manufacturing facilities were consolidated in Chillicothe, Ohio, and the company headquarters was moved there from New Kensington in 1967.

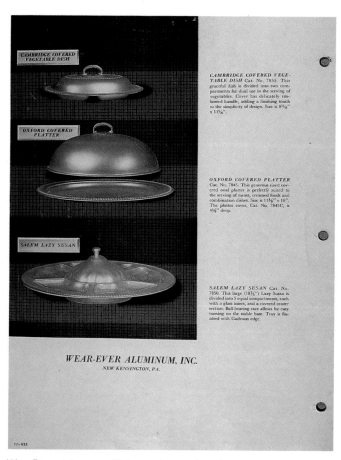

Wear-Ever catalog page. *All vintage photos and catalog illustrations are courtesy Alcoa.*

Kensington

Always seeking new markets, Alcoa decided to expand from purely utilitarian cookware and utensils into a more decorative giftware line, so Kensington Ware was introduced in the Fall of 1934. An announcement in the October 15 edition of Alcoa's company paper, *The Alcoa News*, featured a photograph of the Laurel vase with its stylized brass accents, and described the style as a modern interpretation of Empire. Contemporary critics usually referred to this thirties style as "classic modern," because it had a classical Greek flavor and because it was a modern classic in its own right. The November 12, 1934 issue of *The Alcoa News* featured a more detailed article on Kensington en-

titled "Artist and Artisan Join in Presenting a New Line in a New Metal," which was in response to the many requests for more information. It was described as a new aluminum alloy that produces a whiter color than pure aluminum. New additions to the line were added almost immediately—Compass Plate, tea and coffee services with cherry wood handles, Radcliffe and Chelsea Serving Trays, Wiltshire Flower Dish, Dover Bread Tray, Dorchester Double Serving Dish, Clipper Ship Serving Tray, Coldchester Julep Tumblers, and Piccadilly Smoking Accessories—and were listed in the November 11, 1935 issue in "Timely Aluminum Gift Suggestions for the Christmas Season."

Page from October 15, 1934 issue of *Alcoa News*.

Page from November 12, 1934 issue of *Alcoa News*.

Page from November 11, 1935 issue of *Alcoa News*.

The following is a reprint of a typescript for a promotional piece used by Alcoa sales people entitled "The Story of Kensington" and dated October 10, 1935:

THE METAL

The world is always waiting for something new and different. Gifts made from gold and silver, copper, pewter and bronze have been handed down from generation to generation. Now we introduce a new metal...Kensington.

Kensington metal is an alloy of aluminum, made possible by discoveries of the Aluminum Research Laboratories.

Kensington metal has most of the advantages of silver, pewter, and chromium, yet it has none of their disadvantages. It will not tarnish or stain, but retains its soft, silvery, original lustre. This obviates the necessity of constant cleaning. A mild soap and warm water is all that is necessary, however any ordinary silver polish can be used on Kensington metal without harming the surface.

Kensington metal will not finger print easily, a factor which is common to shiny metal surfaces such as pewter and chrome. Of course it is possible to finger mark the surface with the slightest amount of grease on the finger tips, but under similar circumstances even china will show such marks.

It has a hard surface which is protection against normal wear and scuffing. As there is no superficial plating, there will be no peeling.

DESIGN

Having developed a beautiful metal exceptionally suitable for gift pieces, we were confronted with the problem of choosing a designer. Named by *Fortune* magazine as one of the ten foremost commercial designers in the United States, Lurelle Guild, who was already responsible for two well known gift lines, was our choice.

Against the silvery background of Kensington metal, Mr. Guild has cleverly contrasted exquisite mounts of old brass. This marriage of two metals is in accordance with a new trend in the finest jewelry fields—for example, the combination of platinum and gold, and silver with gold. In several Kensington pieces crystal is used with restraint, while others have no decoration; the beauty of the design and of the metal alone make them equally attractive.

In his designs Mr. Guild has not tied Kensington pieces to any one period. Each piece will go well with the primitive early American as well as complimenting the sophisticated modern. The beauty of Kensington is not merely style deep—it is not something to be bought today and discarded tomorrow, but each piece has an heirloom quality that should last through several generations. Kensington metal engraves beautifully and easily. An engraved piece should always be a part of your display.

FIVE GROUPS

Kensington pieces can be catalogued in the following five groups:

Table and Service items for both formal and informal occasions.

Decorative pieces that sparkle with originality and have a most interesting variety of uses.

Drinking appurtenances having that look of tomorrow that is demanded by the discriminating host.

Smoker's articles that are solidly practical and novel in competition.

Desk accessories which happily wed the functional with the decorative.

DEALER ACCEPTANCE

The first showings of Kensington Giftware were at the Chicago Gift Show in July 1934, and the New York Gift Show in August 1934. Over 4,000 buyers attended these shows, and few left without stopping to see the new Kensington gift line. They were all greatly impressed, and predicted a marvelous reception for our line from the public, a prediction which now has become a realty. We entered the midwinter gift shows as well as this summer's shows in Chicago, New York and Boston. Response was most gratifying in both instances.

THE TRADEMARK

Ever since the 17th century, metal craftsmen have signed their work with their individual marks. The trade mark on each piece of Kensington is the guarantee of its maker. Each Kensington piece is aptly named to give it a personality that will help to sell it.

This Kensington logo with the familiar script beneath the shield appears on the largest number of pieces.

Typical packaging graphics.

Kensington logo on brass portion of vase; vestiges of black crayon line number are visible.

Kensington logo showing stag head above a shield with the letter "K".

THE COMPANY

Kensington pieces are made by Kensington Incorporated in New Kensington, Pennsylvania, a subsidiary of Aluminum Company of America.

NEW YORK SHOWROOM

A permanent Display Room was opened on the 21st floor of the R.C.A. Building in Radio City , in the fall of 1934. This showroom, also designed by Mr. Guild, was inaugurated at a pre-showing for magazine editorial people. They were unanimous in their praise and reaffirmed their approval by giving Kensington an overwhelming amount of publicity in their magazines throughout the past year. Even more extensive editorial mention will be made this fall and winter.

NATIONAL ADVERTISING

This fall we will advertise Kensington in the following class magazines: *Vogue, House Beautiful, American Home, Atlantic Monthly, Harper's Magazine, House & Garden, Country Life, Spur,* and *Time.*

REPEAT BUSINESS

A new metal—unmatched characteristics, authoritative design—a new decorative note—real craftsmanship—wide variety—broad price range—national advertising—editorial approval—gorgeous display possibilities—a product that dignifies any good store—a line that sells and brings customers back for more—that is what Kensington offers you.

KENSINGTON, INC.
NEW KENSINGTON, PA.
October 10, 1935.

By 1936, items were discontinued and advertised in the October 5 issue of *The Alcoa News* in "A Bargain Sale of Kensington Ware! Discontinued Items Are Offered At Low Prices." These items, being offered to Alcoa employees, included 7103 Yorkshire Covered Cake Tray with cover and hard wood board for use as cheese and cracker server; 7318 Marquee Plate Cover for covering toast or hot cakes and fits Kensington sandwich or service plates; 7315 Zodiac 10" Sandwich Plate in all twelve signs of the Zodiac; 7380 Hyde Park Covered Vegetable Dish; 7201 Mayfair Tea Server with aluminum handle, with or without brass decoration on the side, and brass knob on the cover; 7380-B Hyde Park Serving Dish; 7470 Snack Cracker Jar; 7612-A, B, C, & D Royal Family Ash Trays of 4-1/4" diameter fitted with cast snuffers in the shape of King, Queen, Jack, or Joker; 7613 Burleigh Ash Tray of 5-1/2" diameter with brass decoration; 7615-A, B, C, & D King, Queen, Jack, and Joker small cast figural Snuffers; 7471 Sussex Candy Jar with 5-1/8" diameter top; and 7417 Heather Candy or Nut Dish in 3-1/2" diameter, that can also be used as a punch cup.

New products were designed by Guild and introduced into the line through the 1930s and into the '40s. In 1947 Alcoa reintroduced Kensington aluminum and hardwood furniture, with the wood components made in Jamestown, New York and the aluminum parts made by General Fireproofing in Youngstown, Ohio. Guild designed pieces for living room, dining room, and bedroom such as chairs, tables, consoles, vanity, desks, bed headboards, nightstands, chests, and a glazed cabinet.

In an article about aluminum furniture, Gregory Smith explains that Alcoa had begun to experiment with aluminum furniture as early as 1924, but their primary goal was to entice other furniture manufacturers to use the metal. Alcoa was the only supplier of raw aluminum in the United States and was always looking for new markets. These first examples were rather odd, because instead of using appropriately modernistic, even futuristic, designs for this modern metal, Alcoa made aluminum chairs to look like wood. The styles were totally traditional with upholstery covering most of the metal, and faux wood-grain enamel covering whatever remained exposed. In the 1920s Americans were just too conservative in their tastes to accept anything else. By the early 1930s, however, a few machine age, even Art Deco, designs crept into the line, especially for counter stools.

In 1934, the year they introduced Kensington Ware, Alcoa sold its aluminum furniture subsidiary to General Fireproofing. It wasn't until after World War II that they re-entered the furniture manufacturing business with their Kensington line. Guild's designs were, of course, modern, because at least part of the American public was getting over its love affair with historic style. Alcoa sold over 400,000 of these Kensington and Wear-Ever chairs before they were discontinued in 1952.

In addition to the original Kensington Ware aluminum, they also featured a Bent Glass line in similar shapes and motifs for a short time. Another later item was a plain aluminum series of rather uninspired shapes with shiny surfaces covered in a moiré pattern, aptly named Moiré (both are listed in Appendix A). In contrast to the sleek modern look of Kensington Ware, they even produced an oddly fussy traditional line of imitation silver and silverplate with the required ruffles and ridges of good historic reproductions and bad interpretations. Something for everyone? It is not surprising that the most collectible items remain the original Guild designs from the 1930s, especially larger pieces, those with brass accents, and of course, everything that can be classified as Art Deco.

Page from October 5, 1936 issue of *Alcoa News* showing discontinued items.

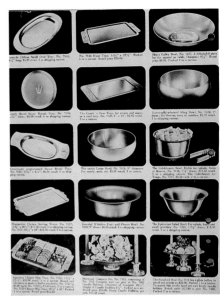

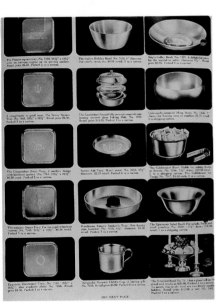

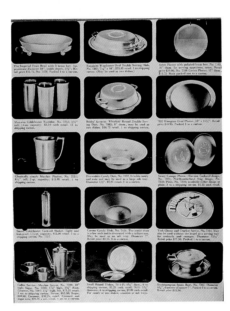

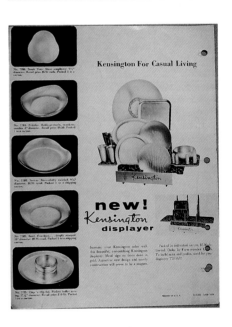

This page, and next page:
Pages from Kensington brochures.

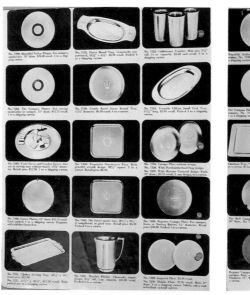

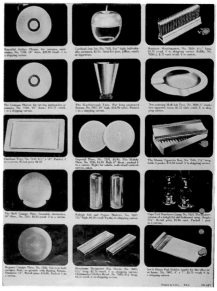

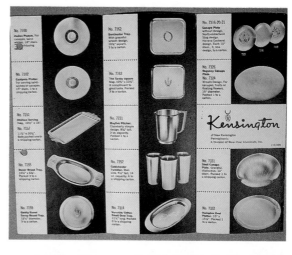

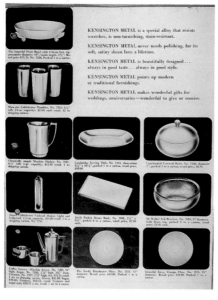

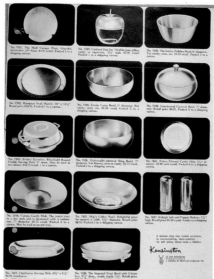

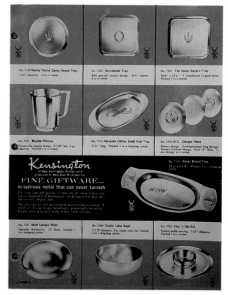

The Designer

Guild began his career in antique furniture, and according to *Forbes* magazine, he wrote an average of five articles each month for various women's magazines and 200 books and pamphlets. Before settling at his design career he traveled around the country and bought antique furniture, wrote articles, and then resold the antiques. His design output was equally prolific, and as one author commented in 1938, "There are few products Lurelle Guild hasn't designed or redesigned. More than a thousand from clocks and thermometers to streamline trains are to his credit" (*Art & Industry*, 228).

Lurelle Van Arsdale Guild was born in Syracuse, New York in 1898. He studied painting at Syracuse University with his future wife and began doing some freelance design for various companies while still a student. After graduating from college, the couple located in New York in 1920, and Guild established himself as an authority on antique furniture. He enjoyed studying seventeenth and eighteenth-century design and crafting methods, and he had a collection of rare books and woodcuts illustrating them. This association with early American (English) design may help to explain the names he later chose for Kensington Ware items, even though machine age is far removed from early American style.

His interest in design and mechanics led him to full-time industrial design work and his own company, Lurelle Guild Associates, with an eight-person staff engaged mostly in model building and product development. He also owned and operated Dale Decorators, a door-to-door decorating service employing more than 100 women who marketed the firm's decorating items such as wallpapers, curtains, and carpeting.

Guild's motto was "mechanical improvement plus eye appeal." Like other industrial designers, he knew that beauty sells, and most of the products he worked on were from the category of art industries—home furnishings and decorative arts. Trained in art and mechanical drawing, Guild usually worked at the drawing board. When he accidently or intentionally came up with a new invention, he would patent it and assign the patent to the manufacturer. Then he charged a fee plus royalties.

Though concerned with a product's outward appearance, Guild knew that skin-deep beauty could be easily pirated, and he stated, "I have found that no matter how good a design, somebody can come along and copy it. But if we make a strong enough patent, so that it supports the design, competing products aren't going to copy the surface appeal as promptly as they will with two products basically similar" (*Art & Industry*, 229). Mechanical improvement are required if the design is to "stick." This means designing from the inside out, starting with crude idea sketches which are made into drawings, then formal working drawings,

full-size plaster or wood models, additional versions of the models if needed, then full size mechanical models, or prototypes, in the actual materials. Since Guild was also very focused on achieving lower manufacturing costs and higher profits, he studied the manufacturer's equipment toward the later stages of the design process. This helped to optimize both the success of the design and its production.

Marketing was at least as important to Guild as design. As one author stated, "Guild is first, last, and always a salesman. His point of departure, therefore, is the market, and he first decides what people want to buy, and what they can pay" (Wickware, 36). He often liked to follow through by conducting his own market research for his designs. Usually he would place samples in retail shops that would carry the product, and he would survey potential customers. Guild is said to have driven a truck filled with refrigerators, including those belonging to the competition, along neighborhood streets. He polled residents

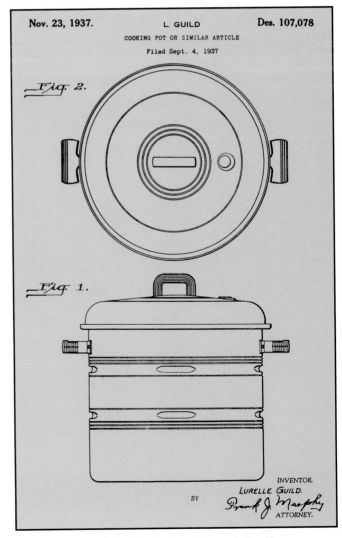

United States Patent Office drawing of cooking pot, designed by Guild for Wearever, filed Sept. 4, 1937 and issued Nov. 23, 1937.

on their preferences and reasons for their choices. This type of grass roots market research technique exemplified Guild's and other designers' attitude of following the public rather than leading it. (Pulos, 268)

In 1936 *Forbes* listed the top five American industrial designers as Henry Dreyfuss, Norman Bel Geddes, Raymond Loewy, Walter Dorwin Teague, and Lurelle Guild. Sometimes known as the gadget man, he styled and shaped thousands of items of hardware, tableware, and kitchenware. New designs for Wear-Ever Aluminum and the new Kensington Ware (with half million dollars in annual sales) accounted for a significant part of his yearly output in the 1930s. The list also includes: Electrolux vacuum cleaner, Norge refrigerator, Thor washer, Williams Oil-O-Matic Corp. boiler-burner, American Locomotive Co. streamlined locomotive, Carrier Engineering Corp. air-conditioning units, General Electric washing and other machines, Montgomery Ward for more washing machines, Ingersoll-Waterbury Co. clocks, Johns-Manville building materials, Underwood-Elliot Fisher typewriters, Weston Electrical Instrument Co. thermometers, Columbia Mills lace curtains and tablecloths, International Silver silver pieces, lamps and other items for Chase Brass & Copper (a list of Guild designs for Chase is in Appendix B), and glassware for Fostoria Glass, such as liqueur glasses with the same laurel leaf motif used for the brass on Kensington Ware.

Exhibit design was another of Guild's many talents, and he designed the Kensington showroom in Rockefeller Center in New York. Notable features included indirect lighting below (rather than above) the display shelves, walls graduated from dark blue-green to white, and a linoleum floor with inlays to carry the color scheme from the walls. He also designed the permanent "museum" at Alcoa's New York offices, in which individual products were presented like works of art. The modern display tables were shapes of extruded aluminum, and the Z-shaped aluminum window treatments eliminated glare while permitting almost full daylight to enter the room at angles.

Rather than just dream up new items for Kensington Ware, Guild responded to the company's perceived needs. In a letter dated Jan. 20, 1938, from H.S. Trump of Kensington Inc. to Guild,

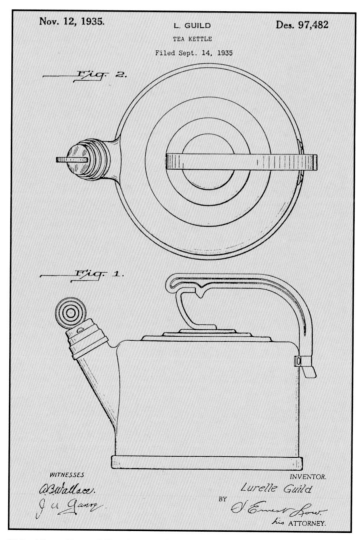

United States Patent Office drawing of tea kettle, designed by Guild for Wearever, filed Sept. 14, 1935 and issued Nov. 12, 1935.

many specific new items were requested. These included: a tray smaller than the Dover bread tray, with different handles; a Clipper Ship design on the $7.50 rectangular tray and a Hunt design on the $5.00 size; a nest of Clipper Ship trays, a pine cone design for the $3.75 canapé plates; a golfer and possibly other sports themes for canapé plates; a smaller pitcher with a "better pouring spout"; a breakfast set with individual coffee pot, cream, and sugar (the bowl of the new baby cup could be used for the sugar and creamer); covered sugar bowl; and an improved spout for the Mayfair Tea Server, because

...when this tea pot is filled it is impossible to pour from it without spilling the contents all over the place. Too it is of little use without a strainer...when a customer pays $15.00 for a tea server it is to serve tea. As the item is turned out now, the tea leaves collect in the spout, for a few seconds no tea comes out the spout, then after giving the pot a shake one gets a mass of leaves and so much tea that it is just too bad. We suggest the spout too at the top be made the same shape as the top of the coffee spout, that is more of a slant instead of the abrupt finish.

Other items requested by Trump included a smaller Marlborough vase; salts and peppers; a simplified double serving dish, without the scalloped edge on the lower pan and without the brass handles; and a larger vegetable dish, by modifying the present one. These very specific needs help to explain why so many slight variations and obvious derivations of Kensington items turn up.

Little is known of Guild's later years; the Archives at Syracuse stop at 1968. In 1979 the American Society of Interior Designer organized an Art Deco exhibit in Pittsburgh and asked Alcoa for examples of Kensington from the 1930s. In a letter from Anna G. Lydon, Alcoa's Records Supervisor, to Guild in 1979, she asked him to help identify the Kensington pieces in their archives. She also mentioned an early exhibit of sixty-one pieces in the RCA Building in New York and asked if he knew of its whereabouts. Guild's reply is worth quoting:

It was not until I received your letter that I started to look into the location of the large important pieces of Kensington which I mentioned to you. After thinking about it further, I remembered that I took many of them to Bermuda for use in my pirate's palace down there, as they do not tarnish like silver. It seems that while I was not on the Island my cook had too much time on his hands. He decided to moonlight and go into the catering business for parties and used the Kensington pieces for serving food. That was no problem for me as he always returned them before I came down to the Island, but recently I learned that he had a heart attack and passed away; with him he took the knowledge of where the pieces are out on loan.

Guild assured Mrs. Lydon that he would locate them when he returned to Bermuda. In the meantime he offered to lend his Kensington pieces that were in Darien, Connecticut (Milestone Village Museum, 1844 Boston Post Road was the letterhead address). The list included "the large hemisphere bowl, large zodiac serving plates with brass mounts, engraved serving plates, covered candy bowl, pitcher, large fruit bowl, large cake server with cover, vase, bread tray, flower bowl, dessert dish, and pepper and salt shakers." He also suggested that she contact the people who were with Kensington at the time they were made—William White, Bob Becker, Dan Stratton, Mrs. Carl Towne, Mrs. Eric Grable, Johnny Hankle, George Hubbard, and Mary Armentrout—most of whom were residing in Pittsburgh (in 1979). He added that if she wanted him to send his Kensington pieces, they should be insured for $700. We can only speculate about the outcome and what happened to the collection when Guild passed away about 1986.

The Product

...ding gifts need not cost a lot

The new, low candlesticks we are seeing on smart tables come in Kensington ware, for about $2.50 each. They are so designed that several can be put together in different ways to form a number of interesting arrangements. A gift all brides will well appreciate

...ly box, which costs
...ections in stacked
...can be used sepa-
...& Copper Co., Inc.

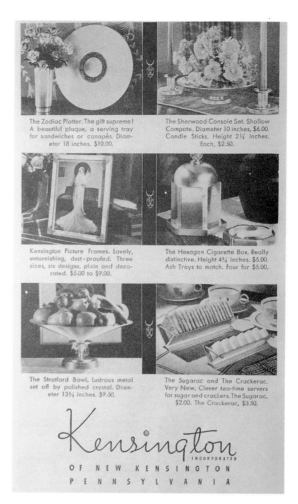

The Zodiac Platter. The gift supreme! A beautiful plaque, a serving tray for sandwiches or canapés. Diameter 18 inches. $10.00.

The Sherwood Console Set. Shallow Compote. Diameter 10 inches, $6.00. Candle Sticks. Height 2¾ inches. Each, $2.50.

Kensington Picture Frames. Lovely, untarnishing, dust-proofed. Three sizes, six designs, plain and decorated. $5.00 to $9.00.

The Hexagon Cigarette Box. Really distinctive. Height 4¾ inches. $5.00. Ash Trays to match. Four for $5.00.

The Stratford Bowl. Lustrous metal set off by polished crystal. Diameter 13⅝ inches. $9.50.

The Sugarac and The Crackerac. Very New. Clever tea-time servers for sugar and crackers. The Sugarac, $2.00. The Crackerac, $3.50.

Kensington
INCORPORATED
OF NEW KENSINGTON
PENNSYLVANIA

Regal Canape Plate

Alcoans can now buy this Hallite combination 2¼-quart cooker and cover and the 7-inch frying pan at real savings. Orders can be placed through the usual employee purchase channels.

Empire Smoker's Tray

SPECIAL FOR...

Wear-Ever Aluminu... service department, is ...lite and Kensington W... more than 60 per cent ... and employees of subsi...

Wear-Ever's spring a... larly suited to solve gift... who is starting houseke... clude a Hallite combi... cover and a 7-inch fryi... for the cooker will fit the... combination is availabl... (Set #3791).

In Kensington Ware,... cial lists the 15-inch di... ideal for canapes, fruit ... and the Empire Smoker... cial with a cork center p...

Form 20223 should ... these special gift item... for the Hallite and K... concerning the employe... been distributed to al... available locally.

Magazine advertisements for Kensington.

Kensington was frequently advertised in magazines, such as *American Home, Atlantic Monthly, Esquire, Harper's, House & Garden, House Beautiful, Time,* and *Vogue* throughout the late 1930s. The following are examples of typical copy accompanying photographs of the wares:

Luxurious beauty is superseding frivolous glitter. Kensington's beauty is rich, substantial, aristocratic. Remember this, also, Kensington metal is a recently discovered alloy of Aluminum which does not tarnish or stain. It keeps its silvery lustre indefinitely. Kensington laughs at time. Lovers of loveliness will want to examine the whole, fascinating family of Kensington gifts, which are displayed by better jewelry and department stores and gift shops. Note to Men: for anniversary gifts, the enduring beauty of Kensington has special significance which will be appreciated. (*American Home, House Beautiful, Vogue,* May, 1937)

Do they open the package saying "How lovely!"...will they prize your present for years and years? Genuine beauty, like that found in Kensington, is always welcome. Discriminating people recognize an authentic elegance which does not stale. And they appreciate this new metal, a recently developed alloy of Aluminum with a lustrous finish. Choose your gifts from the scores of interesting Kensington pieces which you will find at the better stores everywhere. The prices are moderate, ranging from $.50 to $20.00. (*American Home, House & Garden, House Beautiful,* November, 1937)

When your gift must be exactly right, choose Kensington. Scores of beautiful gifts, created by Lurelle Guild and other talented designers, will delight you. Kensington metal is an alloy of Aluminum with charming, silvery lustre.

It keeps its beauty without polishing or special care. To make your gift especially attractive and personal, have it engraved. You will find Kensington gifts at the better department stores, jewelers, and shops everywhere. (*American Home, House & Garden, House Beautiful,* May 1938)

The following catalog of Kensington items covers all of the categories and many of the individual items produced. Most fit nicely into one category, but because of their multi-purpose nature—i.e., tobacco jar or ice bucket, decorative plaque or serving tray—chapter placement is sometimes interpretive and even arbitrary. In most instances, the company issued a line number for an item, and identification is simple. However, for no apparent reason, some numbers were used for more than one item. This will be most apparent in Appendix A.

Some captions include a retail price, which is the price suggested in a company publication, such as a catalog or supplement. Although these prices are the earliest found by the authors, in some cases it is possible that an earlier one exists. Collector prices at the end of the captions reflect the current market values based on antique shops, shows, auctions, and other areas of the retail antique and collectible market. The range takes into account variables, such as geographic region. All items are assumed to be in perfect, or at least near perfect condition. Naturally, some items may be bought and/or sold for more or less than the range, which is provided as a guide and is not intended to set prices. *Neither the authors nor the publisher are responsible for any outcomes based on this guide.* We do, however, wish that all of your Kensington Ware is bought low and sold high. More important, we wish you enjoyment in finding, owning, and learning more about these wonderful examples of American Art Deco, machine-age modern, or just collectible aluminum—whatever label you prefer.

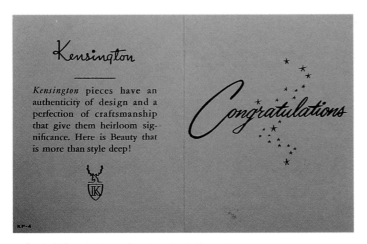

Card of "Congratulations" enclosed with Kensington items.

Inside of "Congratulations" card.

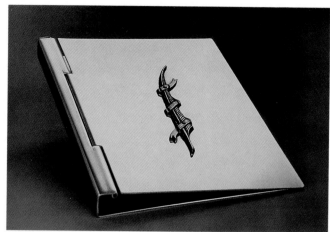

Left:
Original photo of Waverly
Memorandum Book.

Below:
Original photo of
Coldchester Tumblers.

Below:
Original photo of Royal Family
Ash Trays: Jester and Queen.

Original photo of Waverly Pen and Ink Holder.

Original photo of Briarton Covered Bowl and unidentified covered bowl.

Original photo of Sherwood Candleholders.

Original photo of Dorchester Double Serving Dish.

Original photo of Folkestone Bowl.

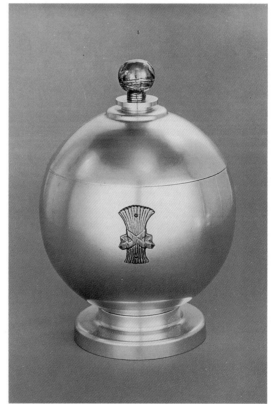

Original photo of round covered container.

Chapter One
Decorative Wares

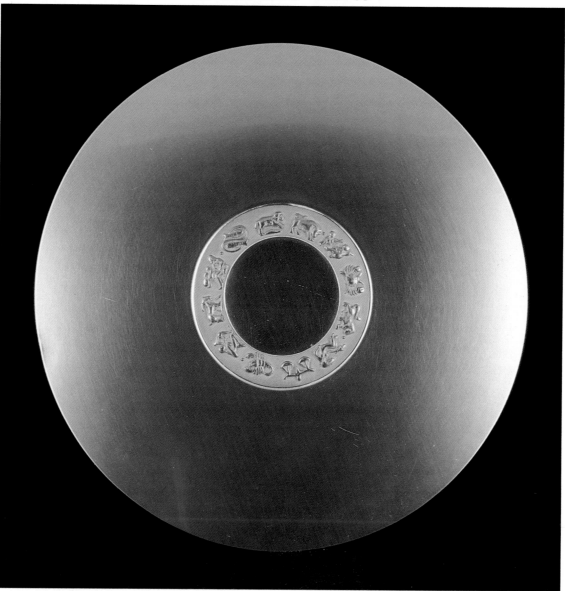

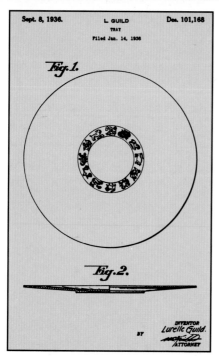

Sept. 8, 1936.   L. GUILD   Des. 101,168
TRAY
Filed Jan. 14, 1936

Fig. 1.

Fig. 2.

INVENTOR
Lurelle Guild.
BY
ATTORNEY

7100 Zodiac Platter, (retail $10.00), with twelve signs of the zodiac on a central circular brass ring, to be used as a canape or sandwich tray or as a decorative plaque, 18" d. $125-150

United States Patent Office drawing of Zodiac Tray, filed by Lurelle Guild Jan 14, 1936 and issued Sept. 8, 1936. The item's introduction preceded the patent application.

Aquarius, the
water carrier.

Pisces, the fish.

Aries, the ram.

Taurus, the bull.

21

Gemini, the twins.

Cancer, the crab.

Leo, the lion

Virgo, the maiden.

Libra, the scales.

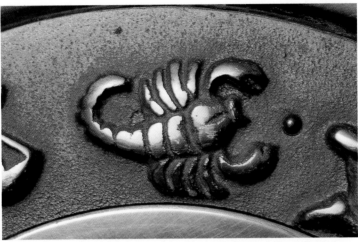

Scorpio, the scorpion.

Sagittarius, the archer.

Capricorn, the goat.

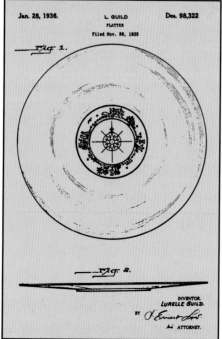

7102 Compass Platter (retail $7.50), with compass points on a central brass band, appeared with the following promotion: "From north, east, south, and west comes enthusiastic acclaim for this new companion to the famous Zodiac Platter. The decorative mount cleverly epitomizes our traditional conceptions of the compass points. A sure-to-be appreciated gift, on any occasion. The Compass serves as luncheon tray, sandwich tray, or canapé tray, and will invariably be used between times as an ever-lustrous decorative plaque." 15" d. (Although the "conceptions" were "traditional," the tray was by any definition modern.) $100-125

United States Patent Office drawing of Compass Platter, designed by Guild, filed Nov. 26, 1935 and issued Jan 28, 1936.

Detail of North, with polar bear and sailing ship.

Detail of East, with pagoda.

Detail of South, with sun and palm trees.

Detail of West, with Amerindian and tepee.

7427 Ship's Galley Bowl (retail $4.50), incised with sailing ship surrounded by eight stars on a domed center, a decorative but also useful piece, 9-1/4" d. $25-30

Detail of ship.

United States Patent Office drawing of Ship's Galley Bowl, designed by Lurelle Guild, filed Sept. 4, 1940 and issued Sept. 9, 1941.

Kensington Picture Frames

"These frames in Kensington metal were first suggested by the need for picture mountings in a metal of distinction which would overcome the disadvantages of rapid tarnishing and high cost....Each design may be had with or without the antique brass mountings at the corners.

Beautiful design is also careful design in Kensington frames. Each back is conveniently fitted with ring and easel, so the picture may be stood or hung. When the picture is fastened to a wall, the inherent lightness of the metal is naturally an advantage. To protect the finest polished surface on which the picture may be stood, the base of the frame-back is curved under the metal molding and covered with soft, durable cloth." (catalog description)

Left: 10130 Picture Frame (retail $6.50), 8" x 10" glass, 5/8" frame, and 7" x 9" picture opening. $60-70

Right: 10030 Picture Frame (retail $5.00), 7" x 9" glass, 5/8" frame, and 6" x 8" picture opening. $50-60

10221 Picture Frame with brass trim (retail $9.00), 10" x 13" glass, 1-1/4" frame, and 7-7/8" x 10-7/8" picture opening. $90-100

10130 Picture Frame displayed horizontally.

Right: Original catalog photograph of 10251 Picture Frame with brass trim (retail $9.00), 9" x 12", 3/4" frame. *Photo courtesy Alcoa* $70-80

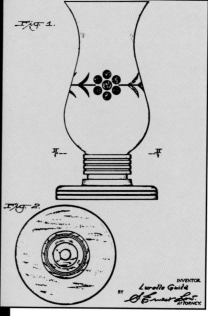

United States Patent Office drawing for Hurricane Lamp, filed Jan. 21, 1941 and issued April 1, 1941.

Left:
7625 Cape Cod Hurricane Lamp (retail $5.75), with glass shade, an example of one of the more traditional designs. 7-3/8" h. $35-60

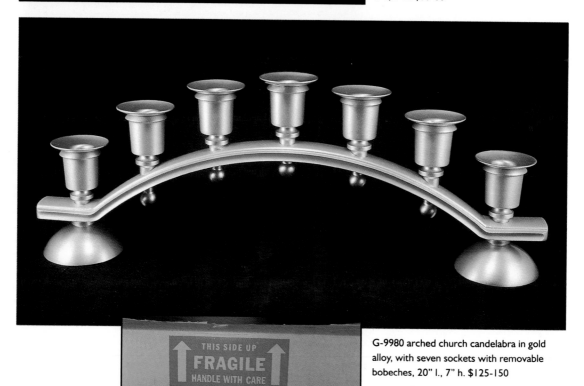

G-9980 arched church candelabra in gold alloy, with seven sockets with removable bobeches, 20" l., 7" h. $125-150

Left:
Packing box from church candelabra, with Kensington script mark, "altar appointments," and stamped "G-9980."

28

Detail of commemorative bowl (7424 Lotus Bowl, 5" d.) produced as a souvenir for the dedication of the Alcoa Building in downtown Pittsburgh in Sept. 1953. $10-15

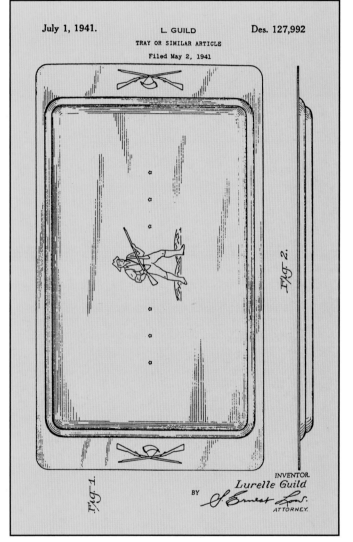

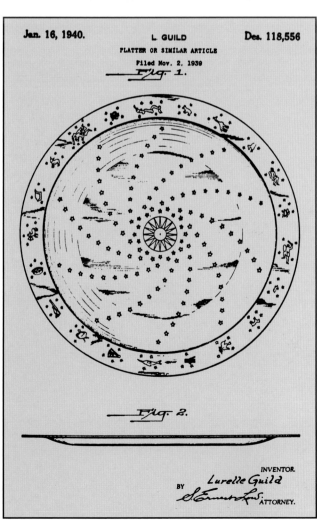

United States Patent Office drawing for a tray, designed by Lurelle Guild, filed May 2, 1941 and issued July 1, 1941. Unknown line number, name, or size, appears to be a Minuteman with musket and powderhorn on rectangular tray. Because metals were in short supply in the early 1940s for anything other than the war effort, many items were temporarily pulled from the line or not produced at all.

Left:
United States Patent Office drawing for 7107 Constellation Platter, by Lurelle Guild, filed Nov. 2, 1939 and issued Jan. 16, 1940. It was introduced for the 1939 Christmas season (retail $12.50); "many pronounce it the most beautiful item in the entire Kensington line. The sun, stars, and constellations of the design are cut sharply into the glowing surface of Kensington. Makes a breath-taking display as a decorative platter, as a centerpiece or in serving canapés or sandwiches." 18" d. $70-90

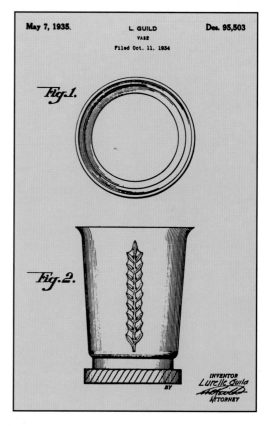

7028 Laurel Vase, part of the original line designed by Lurelle Guild in 1934, with heavy brass foot and brass leaf & branch motif, 9" h. This example has the line number marked in black crayon at the base; it has the shield mark but no name imprinted. $125-150

Top right:
United States Patent Office drawing for Laurel Vase, designed by Lurelle Guild, filed Oct. 11, 1934 and issued May 7, 1935.

Detail of brass laurel motif.

Left: 9" Laurel Vase; right: 7" Laurel Vase. $125-150; $100-125

7031 Marlborough Vase, designed in 1934 (retail $10.00), with weighted brass foot and flared lip, designed to hold long-stemmed flowers requiring some spread, 10" h.; catalog description claims it will "compliment the sophisticated modern." $100-125

United States Patent Office drawing for Marlborough Vase, designed by Lurelle Guild, filed Oct. 11, 1934 and issued May 7, 1935.

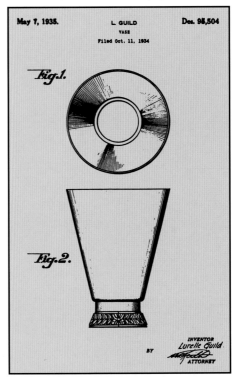

May 7, 1935. L. GUILD Des. 95,504
VASE
Filed Oct. 11, 1934

Fig.1.

Fig.2.

INVENTOR
Lurelle Guild.
BY
ATTORNEY

Below:
Detail of laurel motif on brass foot.

7412 Laurel Bowl, part of the original line designed in 1934, envisioned as a bowl for long-stemmed flowers or fruit, 16-1/4" d., 4-3/8" h.; the laurel motif decorates the underside of the bowl from the foot to the rim; bottom view showing brass trim and base. $175-225

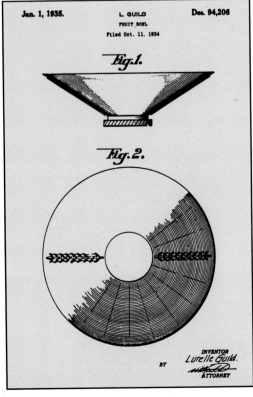

United States Patent Office drawing for Laurel Bowl, designed by Lurelle Guild, filed Oct. 11, 1934 and issued Jan. 1, 1935.

7030 Sherwood Vase, designed in 1934 (retail $4.75),with slightly belled lip and ridged base, 9-1/2" h.; described as "narrow and tall it is intended for medium to long-stemmed flowers. Unadorned, the soft lustre of Kensington metal is sufficient unto itself..." $40-50

United States Patent Office drawing for Sherwood Vase, designed by Lurelle Guild, filed Oct. 11, 1934 and issued May 7, 1935.

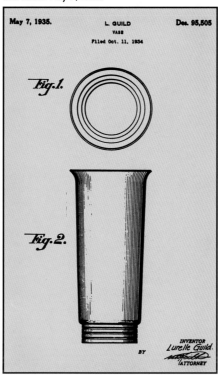

7032 Kingston Vase (retail $5.00), for medium-stemmed flowers, with flared pedestal foot, 6-1/4" h. $25-30

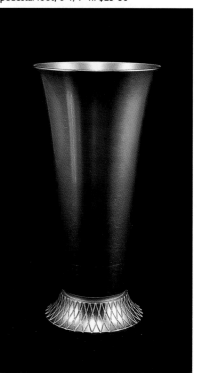

Opposite:
7029 Gainsborough Vase (retail $5.00) with cobalt blue glass insert resting in Kensington base with ribbed pedestal, 9" h. $100-125

Above:
Detail of cross-hatching on base.

7413 Folkestone Bowl, designed in 1934 (retail $8.50), with brass accent trim, mounted on three glass balls resting on aluminum base, 7-3/8" h., introduced for "small fruits and...sizable flower arrangements...and also makes a clever flower pot holder." This is one of only three designs incorporating glass balls. $125-150

Opposite:
7250 Coldchester Wine Cooler, designed c. 1935 (retail $20.00), with brass band with grapes motif supporting movable handles in the form of brass grape clusters, 7-quart capacity, 10-1/4" h., 8-1/4" d.; the fourth book of Kensington gifts states: "It is being said that the final criterion of the accomplished hostess is the correctness of her wine service. For those wines which must be iced, The Coldchester adds the last touch of finesse...It's the right size for a quart bottle of wine." $250-275

Detail of brass applique.

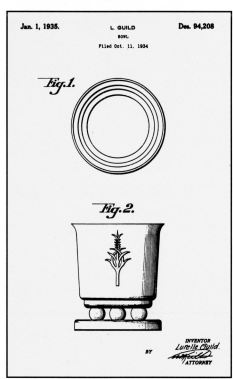

United States Patent Office drawing for Folkestone Bowl, designed by Lurelle Guild, filed Oct. 11, 1934 and issued Jan. 1, 1935.

Detail of brass band with grapes.

Detail of one handle.

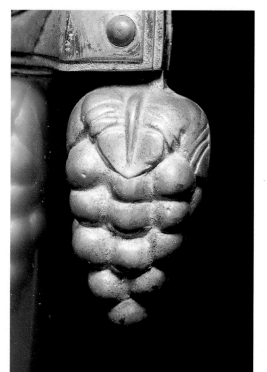

Detail of different section of brass band.

United States Patent Office drawing of wine cooler, designed by Lurelle Guild, filed April 20, 1935 and issued Jan. 28, 1936.

Opposite:
Bulbous covered container, with brass finial and decoration of a bundle of wheat sheaves tied with a ribbon, 10-1/2" h. $175-200

38

Detail of brass knob.

Detail of wheat.

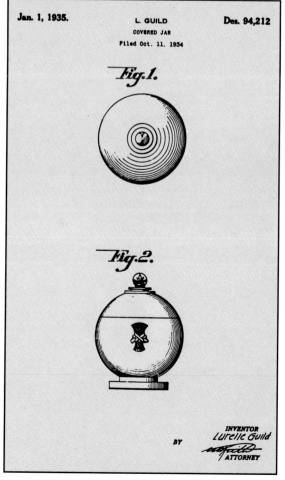

United States Patent Office drawing of covered jar, designed by Lurelle Guild, filed Oct. 11, 1934 and issued Jan. 1, 1935.

Front: pair of 7400 Vanity Fair Candle Holders, designed c. 1935 (retail $3.00), with angular brass trim decorating the base, used to flank some of the other brass-trimmed items, and therefore good sellers, 4" h. $60-80 pair

United States Patent Office drawing of Vanity Fair Candle Holders, designed by Lurelle Guild, filed April 20, 1935 and issued Dec. 10, 1935.

7401 Vanity Fair Tall Candle Holders (retail $3.75), with same brass trim as the shorter version 7" h. $80-100 pair

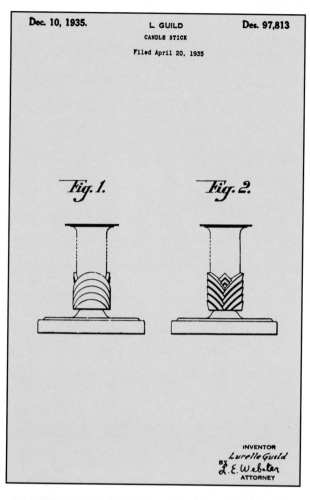

Dec. 10, 1935.   L. GUILD   Des. 97,813
                       CANDLE STICK
                   Filed April 20, 1935

Fig. 1.   Fig. 2.

INVENTOR
Lurelle Guild
BY
L. E. Webster
ATTORNEY

Detail of Vanity Fair trim.

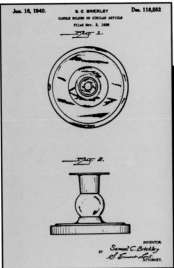

United States Patent Office drawing of Stratford Candle Holders, designed by Samuel C. Brickley, filed Nov. 3, 1939 and issued Jan. 16, 1940.

7405 Stratford Candle Holders, designed by Samuel C. Brickley, c. 1939, (retail $2.50), with crystal ball between the socket and the stepped base, 3-1/4" h. Guild had designed and patented the other two crystal ball items five years earlier. $70-90

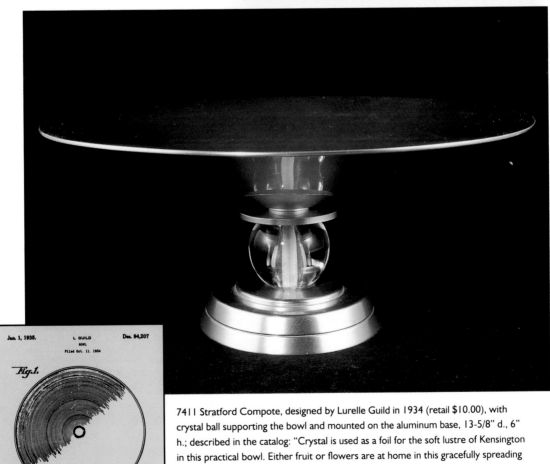

7411 Stratford Compote, designed by Lurelle Guild in 1934 (retail $10.00), with crystal ball supporting the bowl and mounted on the aluminum base, 13-5/8" d., 6" h.; described in the catalog: "Crystal is used as a foil for the soft lustre of Kensington in this practical bowl. Either fruit or flowers are at home in this gracefully spreading piece. Even empty, it is a most satisfying decoration for the buffet or side table." $150-200

Left:
United States Patent Office drawing of Stratford Compote, designed by Lurelle Guild, filed Oct. 11, 1934 and issued Jan. 1, 1935.

7402 Sherwood Candle Holder, designed by Lurelle Guild c. 1937 (retail $2.50), with ribbing on the stick and base, and a wide lip for catching wax drips, 2-1/4" h. $30-40 pair

United States Patent Office drawing of Sherwood Candle Holder, designed by Lurelle Guild, filed Jan. 16, 1937 and issued April 6, 1937.

7422 Devonshire Candy Dish (retail $2.50), coordinates with the Sherwood Candle Holder; first appeared in the fifth catalog as: "Exquisite loveliness of proportion makes The Devonshire a delight to the eye. It holds candy and nuts, and does the other little tasks for which a small odd dish is so useful. It can serve as a large ash tray. Remember that the metal is Kensington, which stays bright and requires almost no care. It follows in design the popular pattern of The Berkeley Bowl." 5-1/2" d. $15-20

Another view of 7402 Sherwood Candle Holders with 7422 Devonshire Candy Dish.

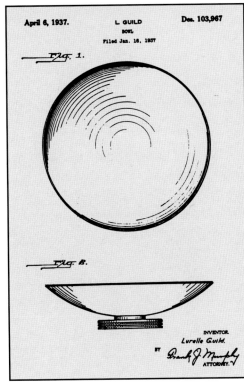

7415 Sherwood Compote, designed by Guild c. 1937 (retail $6.00), which was meant to hold flowers, fruit, or candy and accompany matching candlesticks for a modest console set, 10-1/4" d., 2-3/4" h. $35-45

United States Patent Office drawing of Sherwood Compote, designed by Lurelle Guild, filed Jan. 16, 1937 and issued April 6, 1937.

7409 Hemisphere Bowl (retail $7.50) with ribbing around the top, used for fruit, salad, or punch, 10" d., 5-1/8" h. $50-60

7259 Epicurean Salad Bowl (retail $8.50) with brass foot, described in a June 1936 catalog supplement: "Epicurus—the philosopher who delighted in the pleasures of the senses—would be twice tempted could he see this stunning salad bowl. For the eye, its lines have the balance and charm characteristic of Kensington . For the palate, it promises more pleasures. It will find additional uses when there is occasion to mix a small punch, or to achieve an arrangement of flowers in profuse naturalness. Salad flavoring and natural food acids cannot dim its lustre." 11-5/8" d. $125-150

Detail of brass foot.

7260 Round Table Punch Bowl, designed in 1934 by Guild, with brass foot, six-quart capacity, also suggested for use as a buffet salad bowl, 12-1/4" d., 6-1/4" h. $135-160

Center left:
United States Patent Office drawing of Round Table Punch Bowl, by Lurelle Guild, filed Oct. 11, 1934 and issued Jan. 1, 1935.

Above:
Detail of brass foot.

Detail of brass foot.

7428 Imperial fruit bowl (retail $13.50), resting on three brass winged feet, 9-7/8" d., 2-1/2" h. $75-100

Right: 7246 Coldchester Bowl (retail $5.00), originally introduced as an ice bowl, also used to display flowers or serve salad or fruit, 7-7/8" d. $60-75

Left: 7490 Thistleton covered bowl (retail $3.75), shown without lid, for bon-bons, nuts, or powder, 4-3/4" h. $30-35

Detail of brass knob.

7703 Bowl is the name of this biomorphic shaped serving piece, 10" d. $25-35

Top: 7430 Continental Covered Bowl (retail $8.95), with brass finial or knob, for candy or nuts, 7" d. $35-45

Bottom: 7426 Ming Bowl (retail $5.75), for flowers, nuts, or candy, 7" d. $25-35

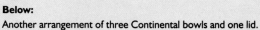

Detail of brass handle with dove and branch motif.

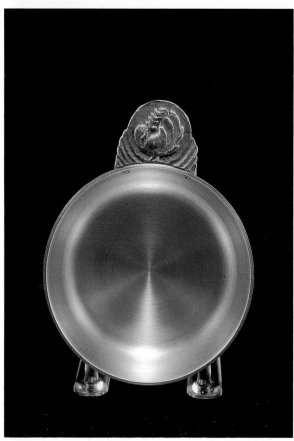

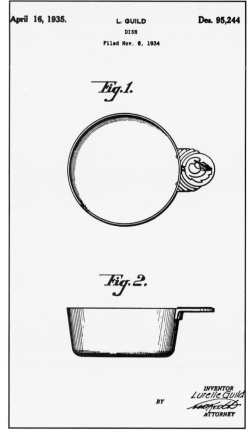

7417 Heather Dish (retail $2.00), with brass handle, for candy and nuts or, as suggested in 1935, for use as a wine taster, punch cup, or deep ashtray, 3-1/2" d. $25-35

United States Patent Office drawing of Heather Dish, by Lurelle Guild, filed Nov. 6, 1934 and issued April 16, 1935.

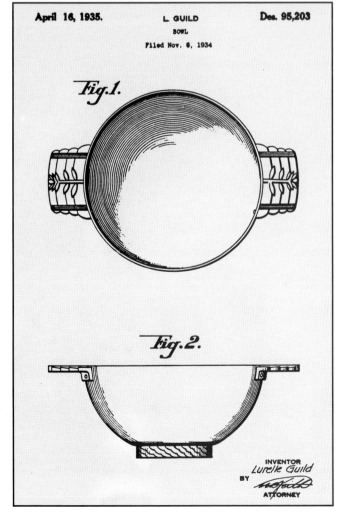

April 16, 1935. L. GUILD Des. 95,203

BOWL

Filed Nov. 6, 1934

Fig.1.

Fig.2.

INVENTOR
Lurelle Guild

BY

ATTORNEY

Above:
7418 Thistle Dish (retail $3.75), with two brass handles and brass base ring, for bon-bons and nuts or "in combination with 7100 Zodiac Platter for pickles, mayonnaise, sauce. Ice bowl. Iced fruit or seafood cocktails. Short stemmed flowers such as pansies, violets, etc." 4-3/4" d. $35-45

United States Patent Office drawing of Thistle Dish, by Lurelle Guild, filed Nov. 6, 1934 and issued April 16, 1935.

Detail of brass handle.

Detail of brass base.

7419 Briar Bon-bon Dish (retail $4.50), described for same use as Thistle Dish, plus as a small fruit bowl for cherries and grapes, 6-1/2" d. $40-50

United States Patent Office drawing of bowl (Briar Dish), by Lurelle Guild, filed Nov. 6, 1934 and issued April 16, 1935.

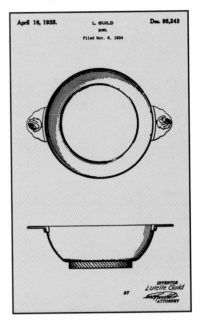

Detail of Briar brass handle with dove and branch, a variation of Heather Dish handle.

7491 Briarton Covered Bowl (retail $5.00), with brass finial, for use as
bon-bon dish, for nuts, small fruit, or as a powder jar, 6-1/2" d. $35-45

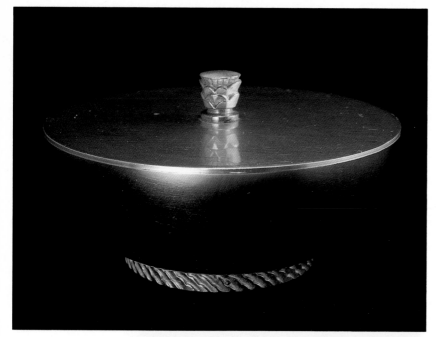

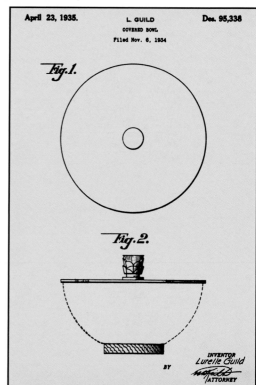

Above:
Covered bowl, with brass
finial and base ring,
illustrated in Nov. 1936
issue of *American Home* as
a first course server,
probably for a cold dish,
3-3/4" h. & d. $35-40

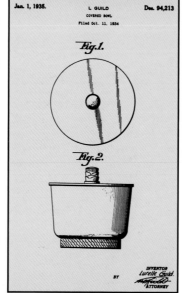

Right:
United States Patent
Office drawing of covered
bowl, by Lurelle Guild,
filed Oct. 11, 1934 and
issued Jan. 1, 1935.

Above:
Detail of brass finial on covered bowl and Briarton Covered
Bowl.

Top right:
United States Patent Office drawing of Briarton Covered Bowl,
by Lurelle Guild, filed Nov. 6, 1934 and issued April 23, 1935.

Opposite:
7472 Nottingham Covered Jar (retail $5.00), with catalog
description: "Uses? Who doesn't need jars for whatever? As a
mate to the severely beautiful lines of the Wiltshire Bowl this
makes a striking combination. Capacious enough for a whole
evening of munching-while-you-read! Men, particularly, like the
lines of this jar." 5-1/2" h., 5" d. $60-70

7410 Wiltshire Bowl (retail $4.00) for short-stemmed flowers or fruit, 3" h., 9" d. $30-40

Opposite top:
Right: 7429 Holiday Bowl (retail $4.95), designed for candy, nuts, etc., 6" d. $25-35
Left: 7424 Lotus Bowl (retail $3.50), with same use as Holiday, 5" d. $20-30

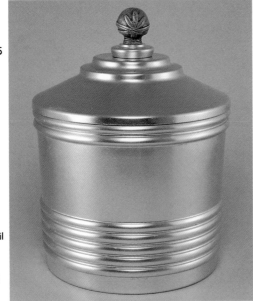

7470 Snack Cracker Jar (retail $7.00), covered canister has banded trim and brass finial; production was discontinued in 1936; 9" h. $65-85

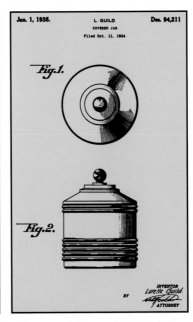

United States Patent Office drawing of covered jar, by Lurelle Guild, filed Oct 11, 1934 and issued Jan 1, 1935.

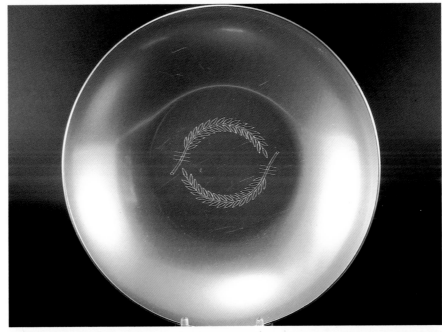

Right:
7432 Golden Dorchester Bowl with wreath design, 9-1/4" h. $15-25

Below:
Detail of laurel wreath, also appearing on other pieces.

59

7381 Dorchester Double Serving Dish (retail $8.50), with brass handles, scalloped bottom, and wreath decorating the lid, 10" d. $50-60

United States Patent Office drawing of Dorchester Double Serving Dish, by Lurelle Guild, filed Nov. 26, 1935 and issued Jan 28, 1936.

Detail of brass handle.

7384 Canterbury covered casserole (retail $15.00), with brass finial, large enough to hold glass casserole included in purchase price, 9-5/8" d. $45-65

Detail of Canterbury finial.

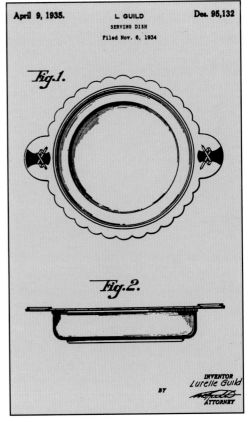

United States Patent Office drawing of serving dish (Hyde Park) with scalloped edge, by Lurelle Guild, filed Nov. 6, 1934 and issued April 9, 1935. The 7380-B Hyde Park Serving Dish (retail $6.50) is without a cover but has brass trim on the handles; it was discontinued in 1936.

7383 Whitfield Double Serving Dish (retail $8.50), with decorative embossing on the lid and handles, this round covered bowl can be used as two separate serving pieces, 9-1/4" d. $25-35

United States Patent Office drawing of Whitfield Double Serving Dish, by Lurelle Guild, filed Feb. 28, 1940 and issued April 16, 1940.

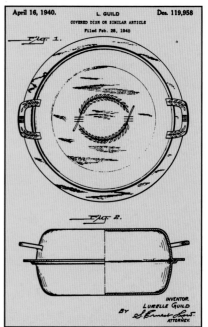

Detail of laurel wreath on handle.

7382 Winchester Double Serving Dish (retail $12.50), an oval covered dish that can double serving capacity when each piece is used individually; like the Whitfield, the lid and handles are embossed, 11-1/4" l. $25-35

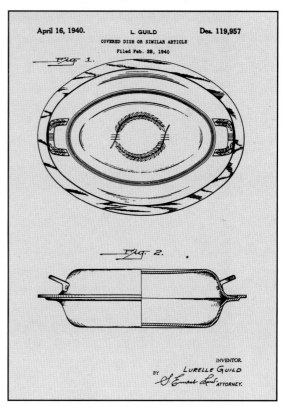

United States Patent Office drawing of Winchester Double Serving Dish, by Lurelle Guild, filed Feb. 28, 1940 and issued April 16, 1940.

Left: 7382-B Manchester Serving Dish (retail $6.50); when the bottom of the 7382 Winchester (shown here separated) is sold alone, it becomes the Manchester, free of decoration, 11-1/4" l. $15-25

7318 Marquee Plate Cover (retail $4.00), fits sandwich and serving plates, has round brass knob, and was discontinued in late 1936, 6-7/8" d. $45-50

Detail of brass finial (knob).

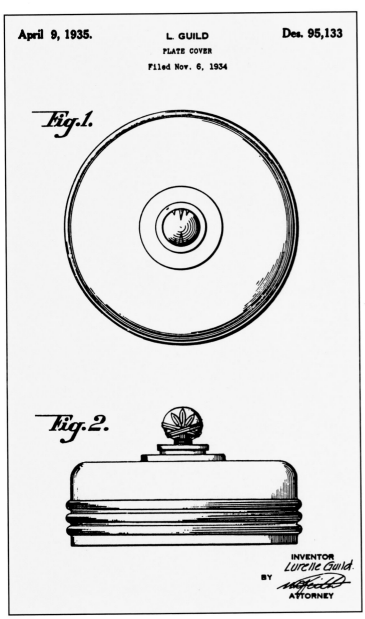

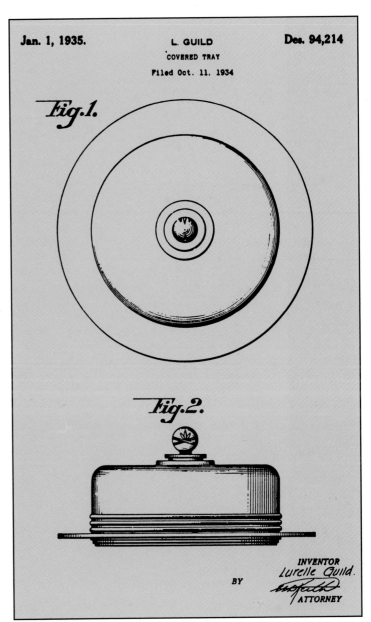

United States Patent Office drawing of Marquee Plate Cover, by Lurelle Guild, filed Nov. 6, 1934 and issued April 9, 1935.

United States Patent Office drawing of covered tray, by Lurelle Guild, filed Oct. 11, 1934 and issued Jan 1, 1935. This appears to be the design for both the 7301 Yorkshire Covered Cake Tray (retail $13.50) and for the 7318 Marquee Plate Cover with serving tray. Both of these items were discontinued in 1936.

7315-D Zodiac Sandwich Plate (retail $2.75), showing astrological sign for Scorpio on brass emblem, in catalog as: ..."sheer beauty. There is a bas-relief mount for each Zodiac sign. What could be smarter than to serve each guest with a plate bearing his own sign? To give the right sign is an incomparable touch of thoughtfulness." 10" d. $25-35

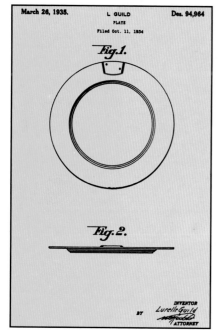

United States Patent Office drawing for Zodiac sandwich and service plates, by Lurelle Guild, filed Oct 11, 1934 and issued March 26, 1935.

Detail of brass Scorpio sign.

Detail of brass Gemini sign.

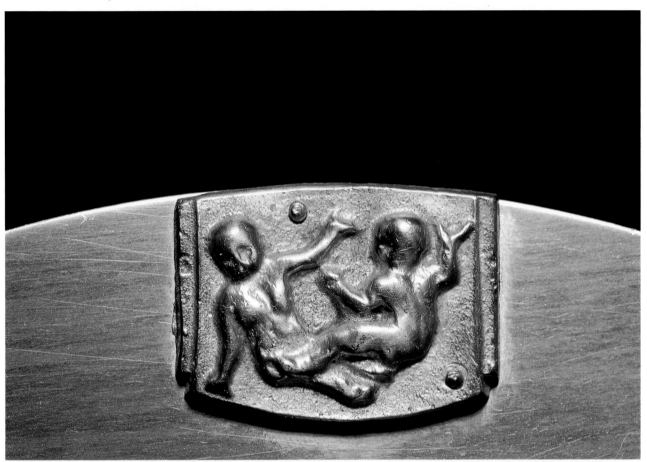

Detail of brass Cancer sign.

Detail of brass Leo sign.

Detail of brass Capricorn sign.

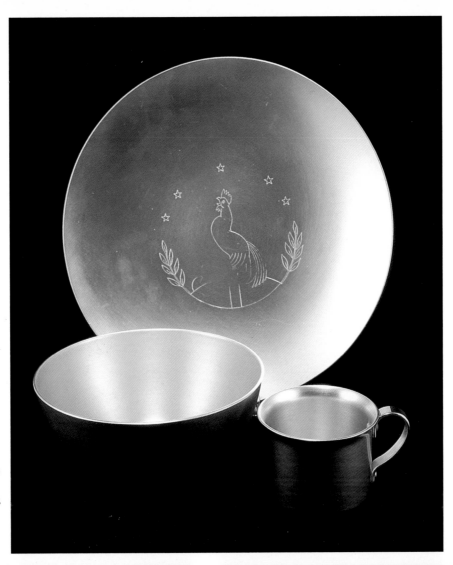

7321 Havana Canapé Plate (retail $3.75), with a crowing cock in the center, is used for serving hors d'oeuvres, sandwiches, or cake, 10" d. $25-30

Above:
7326 Bimini (retail $7.50), introduced as a canapé platter, also touted as a decorative plaque and as a "party companion to the smaller plates, The Havana, The Nassau, and the Northumberland..." 15" d. $40-50

Right:
Detail of cock.

69

7331 Shell Canapé
Plate, in the form of a
shell, with incised
lines, 10" d. $15-20

Original photograph of 7254 Five O'Clock Cocktail and Canapé Server (retail $2.00), described as follows: "Definition of a cocktail: 'Something which while you have a sandwich in one hand, and an hors d'oeuvre in the other, you hold in your *other* hand.' The Five O'Clock Server is designed for coping with such dilemmas. Besides facilitating the service of inevitable cocktail, canapé, and sandwich, it can do triple duty as a card tray, an ash tray, or as a coaster for tall drinks in goblets and glasses." 6" d. *Photo courtesy Alcoa*

Right:
United States Patent Office drawing of Shell Canapé Plate, by Lurelle Guild, filed April 15, 1939 and issued Nov. 21, 1939.

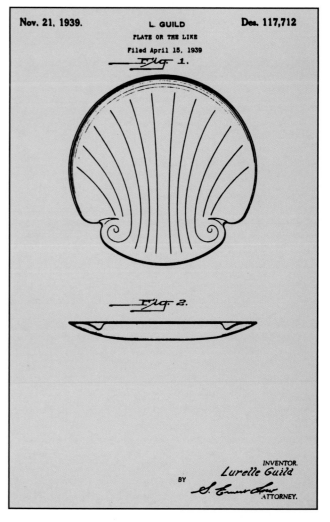

Nov. 21, 1939.   L. GUILD   Des. 117,712
PLATE OR THE LIKE
Filed April 15, 1939
Fig. 1.

Fig. 2.

INVENTOR.
Lurelle Guild
BY
S. Ernest Law
ATTORNEY.

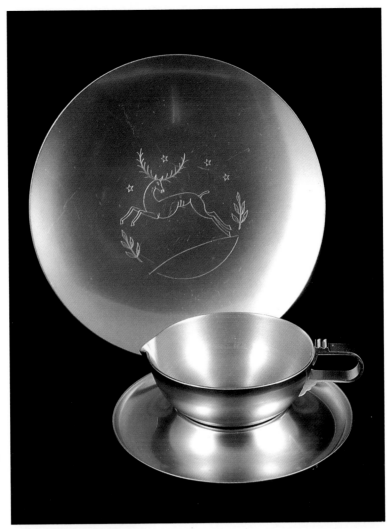

Back: 7320
Northumberland Canapé
Plate (retail $3.75),
depicts frolicking stag
surrounded by stars and
wreath, 10" c. $25-30
Front: 7261 Southampton
Sauce Boat (retail $6.00),
for gravies, salad, or
mayonnaise, 7" saucer,
4-3/4" boat. $35-40

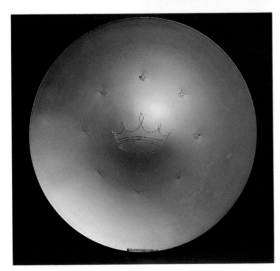

7326 Regency Canapé Plate, also called Regal Canapé Plate,
(retail $10.00) with crown and stars decoration, used for
serving canapés or fruit, 15" d. $35-45

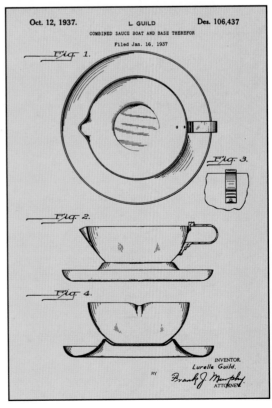

United States Patent Office drawing of Southampton
Sauce Boat, by Lurelle Guild, filed Jan. 16, 1937 and
issued Oct. 12, 1937.

Top: 7159 Savoy Round Tray (retail $5.00), serves sandwiches or canapés, and its flared edge makes it suitable to carry drinks, 12-1/4" d. $25-35

Center: 7301 Vanity Fair Butter Plate (retail $1.75 plain or $2.00 decorated), with brass mount on edge, also used as an individual canapé plate or lemon plate, 5-1/2" d. $20-25

Bottom: 7607 Hostess Ash Trays (retail $.75), 2-5/8" d. $3-5 each

Below:
Detail of brass on Vanity Fair Butter Plate.

Left: 7158 Irvington Tray (retail $3.75), with repoussé linear Art Deco design, for bread or practically anything, 6" x 13". $20-30

Right: 7155 Dover Tray (retail $3.75), common item, with pineapple motif at each end, described in the catalog: "Generously proportioned, The Dover is shaped to accommodate bread or rolls, muffins, or crackers. Reasonably priced, this is a splendid gift piece. Decorations in repoussé. Substantial thickness." 6" x 13". $15-20

Detail of Art Deco design.

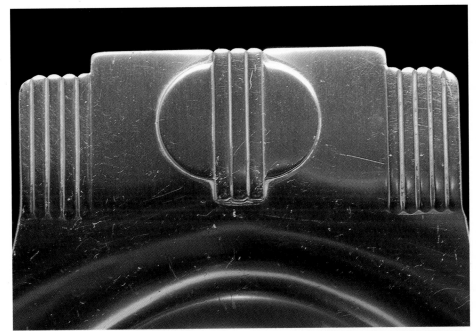

Detail of pineapple.

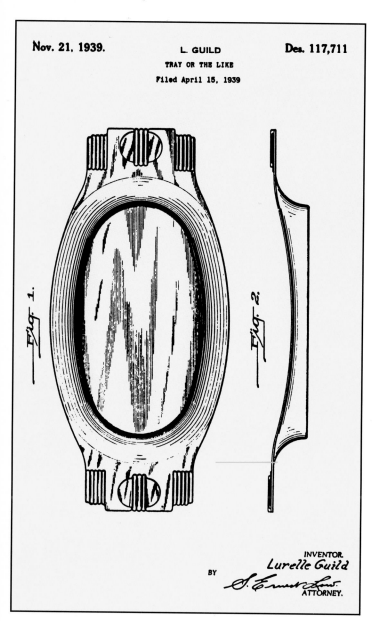

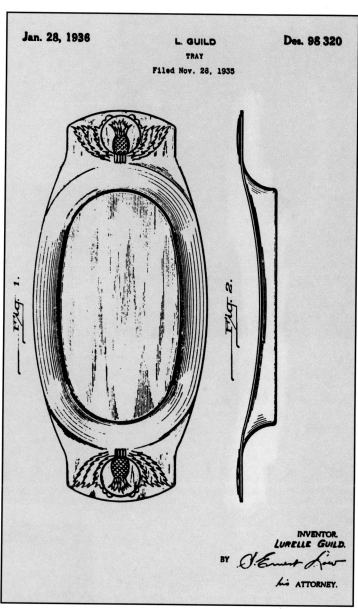

United States Patent Office drawing of Irvington Tray, by Lurelle Guild, filed April 15, 1939 and issued Nov. 21, 1939.

United States Patent Office drawing of Dover Tray, by Lurelle Guild, filed Nov. 26, 1935 and issued Jan. 28, 1936.

Front: 7314 Clifton Celery Dish (retail $3.00), a short-lived piece with central embossed design, 11-1/4" l. $20-30
Back: 7157 Chelsea Serving Tray (retail $17.50), 11-3/4" x 20 1/2'. $35-45

Detail of Art Deco leaf on Chelsea Tray.

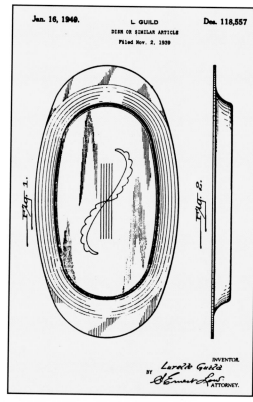

United States Patent Office drawing of Clifton Dish, by Lurelle Guild, filed Nov. 2, 1939 and issued Jan. 16, 1940.

Detail of motif on Clifton Dish.

Back: Golden Dorchester Canapé Plate, round with central wreath decoration, 10" d. $20-35
Front: 7611-B Prince Edward (retail $1.50), with classic medallion design, dished and suitable for mints or as an ashtray, 5-3/8" d. $10-15

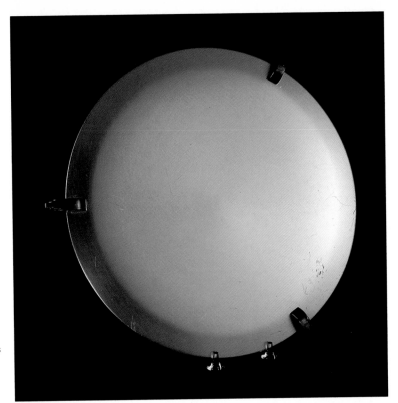

7106 Cortez Footed Platter (retail $7.50), with polished brass Art Deco winged feet, may be used "to make [a] strikingly beautiful centerpiece, or for more practical purposes...to serve hors d'oeuvres, sandwiches, cakes, cheeses, and other delicacies. A gift which the bride welcomes." 12 " d. $75-85

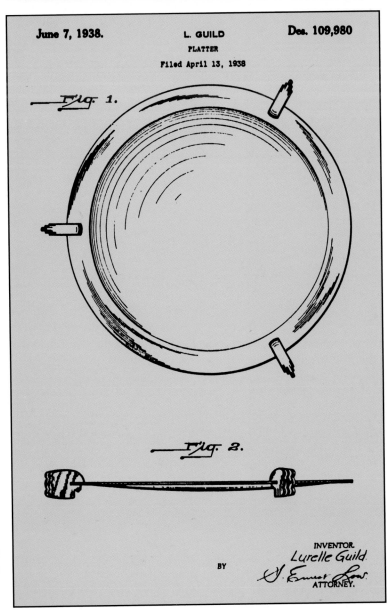

June 7, 1938.

L. GUILD

PLATTER

Filed April 13, 1938

Des. 109,980

Fig. 1.

Fig. 2.

INVENTOR.
Lurelle Guild.

BY

ATTORNEY.

United States Patent Office drawing of Cortez Footed Platter or Aztec Footed Platter (the 16-inch version), by Lurelle Guild, filed April 13, 1938 and issued June 7, 1938.

Detail of brass Art Deco winged foot.

Back: 7332 Hampton Oval Platter (retail $7.50), with repoussé
decoration at both ends, 10-3/4" x 15-3/8". $30-35
Front: 7697 Raleigh Salt and Pepper Shakers. $15-20

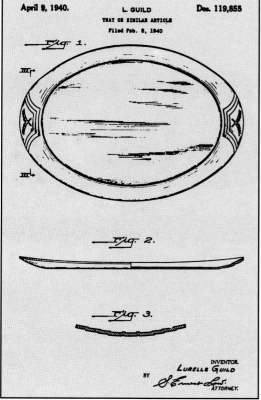

April 9, 1940.

L. GUILD

Des. 119,855

TRAY OR SIMILAR ARTICLE

Filed Feb. 8, 1940

Fig. 1.

Fig. 2.

Fig. 3.

INVENTOR.
LURELLE GUILD
BY
ATTORNEY.

United States Patent Office drawing of Hampton Oval Platter,
by Lurelle Guild, filed Feb. 8, 1940 and issued April 9, 1940.

7165 Queen Tray (retail $6.50), hostess tray for hors d'oeuvres or petit fours, 10-3/4" sq. $20-25

7162 Dorchester Tray (retail $6.50) and 7060 Golden Dorchester Tray (retail $8.95), each dished and decorated with wreath motif, 10-3/4" sq. $20-25 each

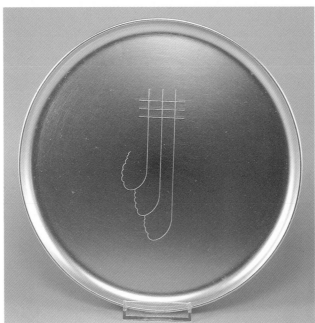

7329 Imperial Plate, decorated with highly stylized linear design, used for cakes or anything else, 8" d. $25-30

United States Patent Office drawing of Raleigh Salt and Pepper Shakers, by Lurelle Guild, filed Nov. 2, 1939 and issued Feb. 13, 1940.

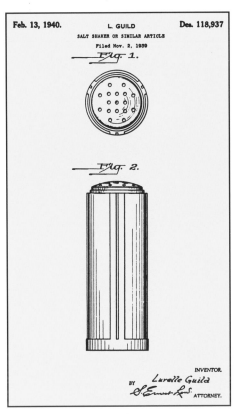

7697 Raleigh Salt and Pepper Shakers (retail $3.75), with vertical ribbing, bottom openings, 2-3/4" h. $15-20

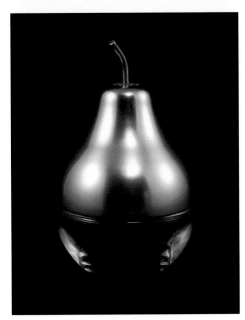

7421 Croydon Jam Jar (retail $3.00), described in the fifth gift catalog as: "A pear-shaped jam jar is authentically Early American, though The Croydon's simple lines are in keeping with modern furnishings. The jar itself is crystal glass, the cover is Kensington metal with a brass stem. There are no sharp corners or crevices to clean. Hostesses will recognize in the Croydon an ideal bridge prize, useful for serving jams, jellies, candies or nut meats. As a thoughtful touch fill it with sweets before giving it." 6" h. $25-35

7423 Cortland Jam Jar (retail $4.00), presented in the catalog as: "Another among the fructiform containers now so popular...Its very shape suggests the cidery tang of apple jelly. Beside its use for jams and jellies, The Cortland may be filled with candy or cigarettes. You can depend on its well-fitted cover to guard the freshness of spicy, preserved ginger. It is just the thing for difficult gift lists, because it is something new for the woman who 'seems to have every-thing.'" 5-1/2" h. $25-35

Croydon and Cortland Jam Jars, examples of the few items of glass and aluminum; not among the most interesting or important designs, because they were trying to appeal to both traditional and modern tastes.

7103 Yorkshire Covered Cake Tray (retail $13.50), available with center hardwood board, also used to serve cheese and crackers, 15-3/8" d. x 6-1/4" h. $175-225

7725 Chip and Dip Set (retail $10.95), perfect for buffet service, 12-1/4" d. base, 4-3/4" d. bowl. $35-45

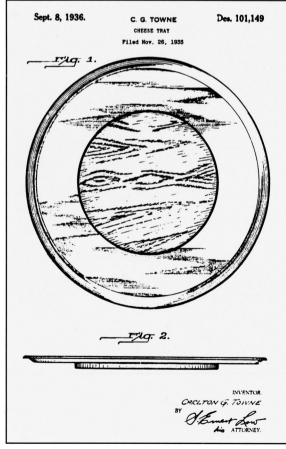

7104 York Cheese & Cracker Service (retail $12.50), designed by Carlton G. Towne, used as a serving tray if the cheese board is removed, 15-5/8" d. $50-60

United States Patent Office drawing of York Cheese & Cracker Service, by Carlton G. Towne, filed Nov. 26, 1935 and issued Sept. 8, 1936.

Back: 7112 Mayfair Round Tray (retail $10.00), with handles of wood with incised design of circles, straight lines, and stylized leaves, 15" d. (also produced in smaller size) $75-100

Front: 7108 Laurel Tray (retail $7.00), with brass handles, (line number marked on the back in crayon), 12" d. $100-125

7110 Laurel Tray (retail $10.00), with brass handles, promoted in gift catalog: "No home can be quite complete without a full budget of round trays, and that's why the Laurel was made. With quiet elegance it holds the coffee or tea service, or a load of tall and cooling drinks. Dainty women, insistent that their every possession reflect exquisite taste, prefer its delicate ornamentation. Many choose to have the full set of three sizes." 18" d. $150-175

Detail of brass handle.

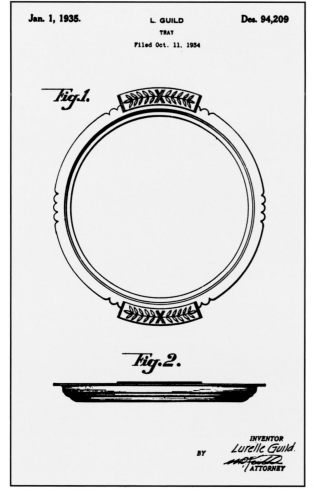

Jan. 1, 1935.   L. GUILD   Des. 94,209
TRAY
Filed Oct. 11, 1934

Fig.1.

Fig.2.

INVENTOR
Lurelle Guild.
BY
ATTORNEY

United States Patent Office drawing of Laurel Tray, by
Lurelle Guild, filed Oct. 11, 1934 and issued Jan. 1, 1935.

In contrast to the modern tray designs were several that were clearly traditional:

Back: 7154 Hunt Tray (retail $7.50), promoted as: "In tune with the zest of life—The Hunt Tray is just the thing for serving party cocktails, afternoon tea, after-dinner coffee. The decorative hounds and hunters are in repoussé. The essential lightness of Kensington gives a unique advantage. Naturally this tray will never stain or tarnish." 11-3/4" x 20-1/2" (also made in 14-3/8" x 22-5/8" size and with optional buffet server) $50-75

Front: 7149 Coach 'N Four Tray (retail $3.50), described: "This smart, convenient little tray... is intended primarily for the sacred cocktail—but the tray finds further usefulness for serving canapés, and as a card tray." 6" x 11". $40-50

Detail of Hunt.

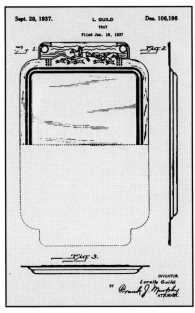

United States Patent Office drawing of Hunt Tray, by Lurelle Guild, filed Jan. 16, 1937 and issued Sept. 28, 1937.

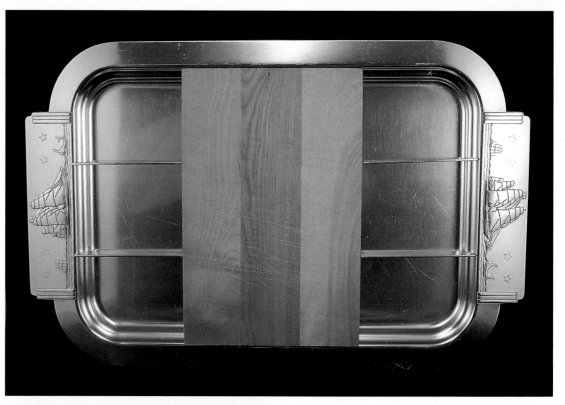

7152 1/2 Clipper Ship Buffet Server (retail $12.50 and $2.50 for dividers), described: "If the appearance of this tray at the buffet supper does not bring a chorus of *ohs* and *ahs*, there are no aesthetes, no gourmets left in the world. Generous board for the bread or crackers, and six sizable compartments for things that make you smack your lips. When the dividers are removed, you have a useful serving tray." 14-3/8" x 22-5/8" (also sold in 10-1/2" x 18" size with cover and dividers forming four compartments). $125-150

Detail of nautical scene.

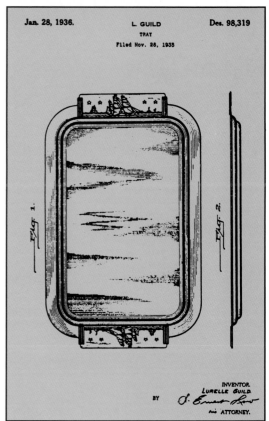

United States Patent Office drawing of Clipper Ship Tray, by Lurelle Guild, filed Nov. 26, 1935 and issued Jan. 28, 1936.

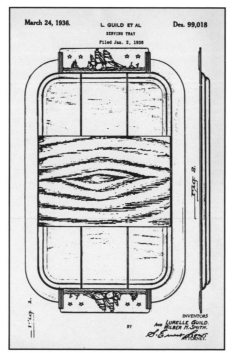

United States Patent Office drawing of Clipper Ship Buffet Server, by Lurelle Guild and Wilber M. Smith, filed Jan. 2, 1936 and issued March 24, 1936.

United States Patent Office drawing of tray, by Lurelle
Guild, filed Oct. 30, 1935 and issued Jan. 28, 1936.

United States Patent Office drawing of bowl, by Lurelle
Guild, filed Feb. 8, 1940 and issued April 30, 1940.

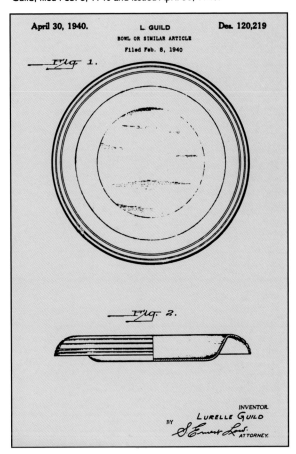

United States Patent Office drawing of bowl, by Lurelle
Guild, filed Jan 21, 1941 and issued March 18, 1941.

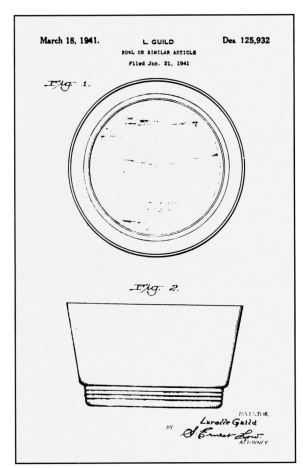

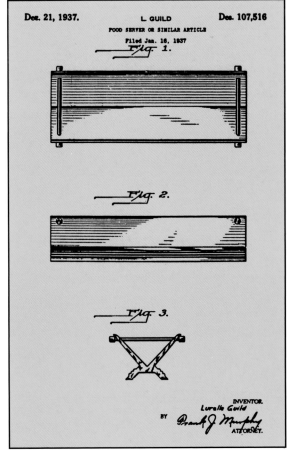

United States Patent Office drawing of food server, by
Lurelle Guild, filed Jan. 16, 1937 and issued Dec. 21, 1937.

Detail of flying gulls motif of Kensington shipping line logo; this decorated several standard Kensington items, which may have been given different line numbers. However, the following shipping line items are identified by shape.

Round serving dish with center handle and small shipping line logo, 10" d. $15-20

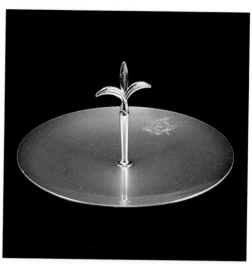

Back: another view of serving dish with central handle; front: dished ashtray or mint dish, 5" d. $5-10

7148 Chatham Tray, featuring a map of the Gulf of Mexico and Caribbean, with shipping logo, 8-1/2" x 14". $30-35

7162 Dorchester Tray with shipping emblem; back tray is
silver, and front tray is gold, 10-3/4" sq. $15-25 each

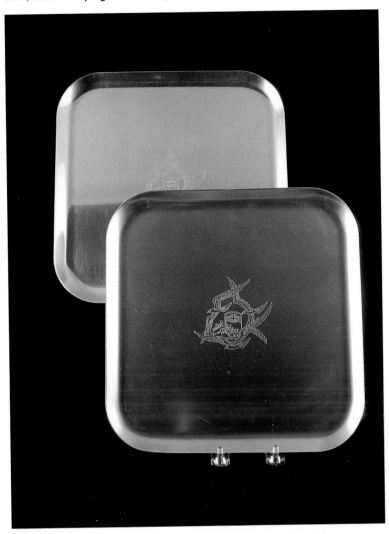

7316-P Claridge Canapé Tray with shipping logo,
10" d. $20-25

Left: 7160 Pioneer Tray (retail $6.50), with antique locomotive decoration;
10-3/4" sq. $20-30
Right: 7330 Malolo Plate (retail $4.95) for salads, petits fours, or cocktails;
features pineapple motif with two leaves; 8" d. $20-30

7231 Mayfair Pitcher (retail $8.75), with brass emblem of dove and branch, catalog description: "Matching The Mayfair Coffee Service and Tea Service, this generously-sized pitcher completes a serving ensemble of true distinction. Mayfair both *looks* and *pours* the part, whether it be used at the buffet supper or to replenish the glasses at the formal dinner. The Mayfair design, classic in its simplicity, is an excellent example of the ability of Kensington to harmonize with any décor. Capacity, two quarts." 8-1/8" h. $60-75

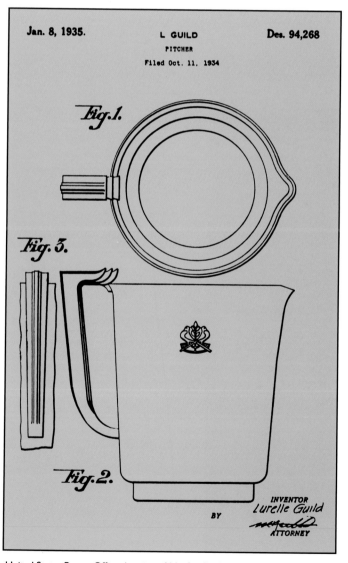

United States Patent Office drawing of Mayfair Pitcher, by Lurelle Guild, filed Oct. 11, 1934 and issued Jan. 8, 1935.

Detail of brass emblem.

Two different handle designs used for the Mayfair Pitcher: left is
thinner and ridged at top; right is thicker with ridges touching the rim.

Next page:
7252 Coldchester Tumbler (retail $2.50), described in catalog: "Frosty tall
ones—juleps with their sprigs of mint—lemonades—iced tea—highballs—
even beer—you could very properly imagine that the soft lustre of
Kensington was created just to glorify the cold summer drink. The graceful
simplicity of line accentuates the intrinsic beauty of the velvet-smooth
texture of Kensington metal." 14-oz. capacity, 5-1/4" h. $7-10

7730 Serving Pitcher (retail $15.95), with ice lip and black
Bakelite handle, two quart capacity, 8-1/4" h. $70-80

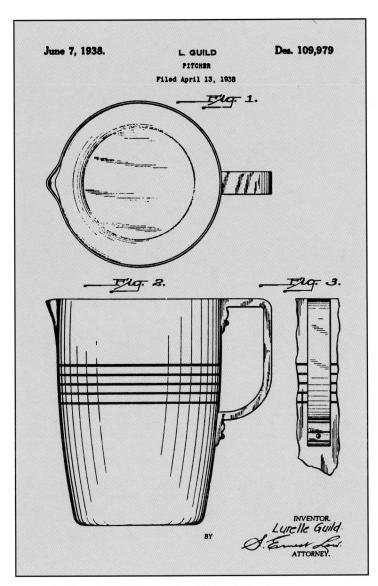

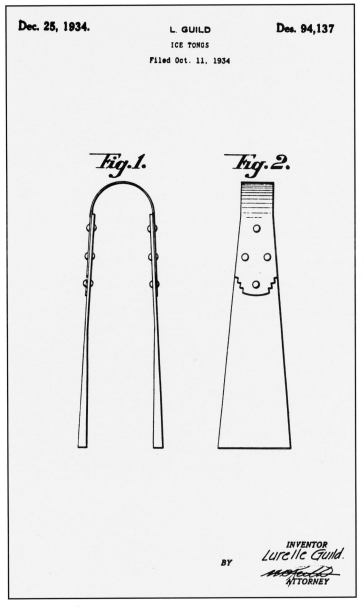

United States Patent Office drawing of Riviera Pitcher, by Lurelle Guild, filed April. 13, 1938 and issued June 7, 1938.

United States Patent Office drawing of tongs, by Lurelle Guild, filed Oct. 11, 1934 and issued Dec. 25, 1934. These 7245 Coldchester Tongs, brass on aluminum, were shown in the 1935 gift catalog but soon replaced by the plainer version without brass.

Opposite:
7232 Riviera Pitcher (retail $10.00), two quart capacity and sleek modern design, 7-1/2" h. (shown with 7471 Sussex Tobacco Jar). $40-50

Left: 7247 Coldchester Ice Tongs (retail $1.75), replaced original tongs. 6-1/8" l. $20-25
Right: 7245 Newport Ice Bowl (retail $3.50), which evidently replaced the original Coldchester Ice Tongs by using the same line number; the ribbing on the lower portion of bowl complements the 7247 Tongs, 6" d. $35-40

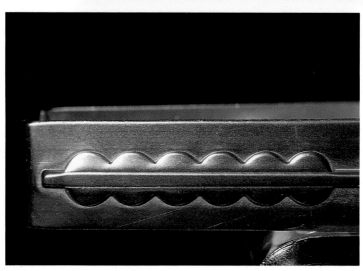

Detail of Art Deco motif on 7247 Ice Tongs.

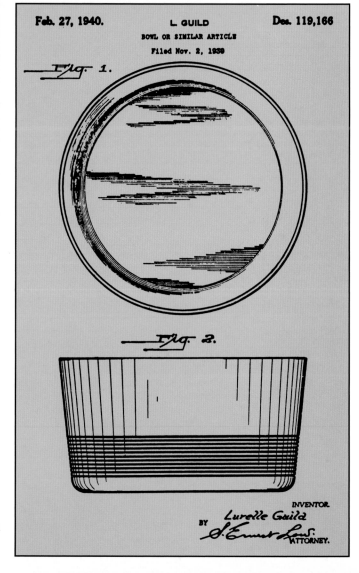

United States Patent Office drawing of Newport Ice Bowl, by Lurelle Guild, filed Nov. 2, 1939 and issued Feb. 27, 1940.

7251 Coldchester Cocktail Shaker (retail $10.00), described in fourth gift catalog: "It is high time for a Cocktail Shaker that is neither bizarre nor mid-Victorian. Coldchester is an aristocrat. It has the distinction that will make its owner proud—and his guests envious! Tarnishing and staining are alike unknown. Cleverly fitted with cork to eliminate possibility of leakage. Substantially made, zealously crafted. Capacity 1-1/2 quarts." 13-1/8" h. $100-125

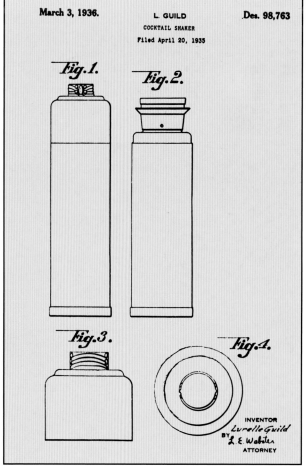

7253 Coldchester Cocktail Cups (retail $1.00 each) were sold in sets of four and advertised with the 7149 Coach 'N Four tray as pictured; 3-oz. cups doubled as individual cigarette holders; 2" h. $10-15 each

United States Patent Office drawing of Coldchester Cocktail Shaker, by Lurelle Guild, filed April 30, 1935 and issued March 3, 1936.

Disassembled shaker items: cork, strainer, base, and lid with brass knob.

Opposite:

7200 Mayfair Coffee and Tea Service, with cream and sugar, was part of the inaugural production. These pieces were all made with aluminum handles, brass finials on the lids, and dove and branch emblems on the fronts. Shown on 7110 - 18" Laurel Tray.

Left: 7201 Mayfair Tea Server (retail $15.00), 6-cup capacity, 7" h. to finial top. $80-90

Back: 7200 Mayfair Coffee Server (retail $20.00), 12-cup capacity, 10" h. $70-80

Front: 7202 Mayfair Creamer (retail $4.00), 3-1/8" h. to handle top. $25-35

Right: 7203 Mayfair Sugar (retail $4.00), 3-1/4" h. $25-35

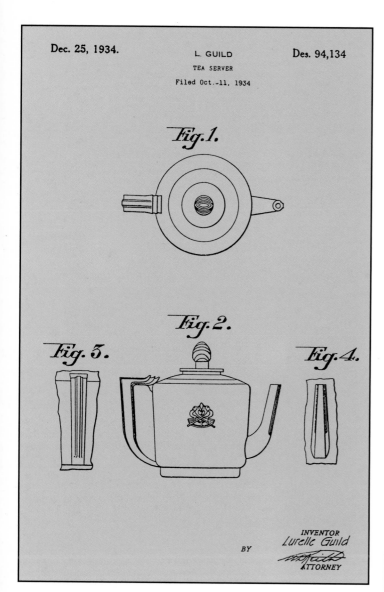

United States Patent Office drawing of Mayfair Tea Server, by Lurelle Guild, filed Oct. 11, 1934 and issued Dec. 25, 1934.

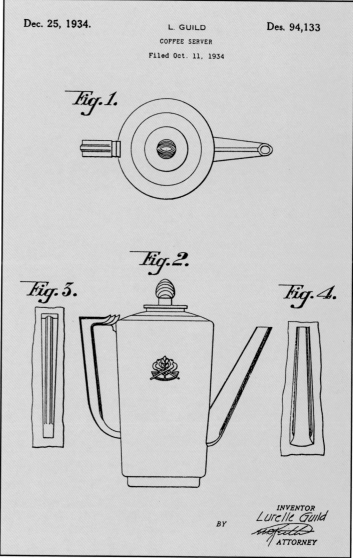

United States Patent Office drawing of Mayfair Coffee Server, by Lurelle Guild, filed Oct. 11, 1934 and issued Dec. 25, 1934.

Closer view of Mayfair Cream and Sugar.

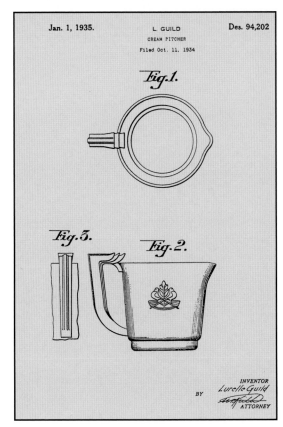

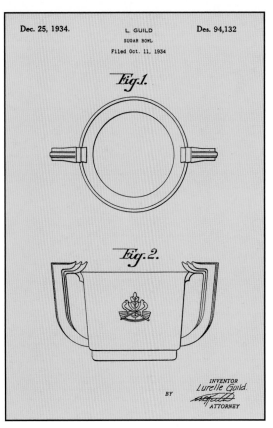

United States Patent Office drawing of Mayfair Cream Pitcher, by Lurelle Guild, filed Oct. 11, 1934 and issued Jan. 1, 1935.

United States Patent Office drawing of Mayfair Sugar Bowl, by Lurelle Guild, filed Oct. 11, 1934 and issued Dec. 25, 1934.

Subtle variation in Tea Server spout: original Mayfair (left pot) had a problematic design, so in 1938 Guild redesigned it with an angle more like that of the Coffee Server (right pot).

Opposite:
In late 1935 the 7100 Mayfair Tea and Coffee Servers, Cream, and Sugar were manufactured with natural cherry wood handles and finials in place of metal, and promoted as: "cherry wood handles and knobs that stay cool." A larger creamer, the 7207 Beverly was introduced in 1939.
Back left: 7201 Mayfair Tea Server, 7" h. $65-75
Back right: 7200 Mayfair Coffee Server, 10" h. $50-60
Front left: 7202 Mayfair Creamer, 3-1/4" h. $15-25
Front right 7203 Mayfair Sugar, 3-1/4" h. $15-25
Shown on 7112 Mayfair Tray with incised design on wooden handles, 15" d. $50-60

Mayfair Coffee Server.

Mayfair Creamer.

Mayfair Sugar, showing wood handles with ridges on top similar to the discontinued aluminum handles.

Comparison of the smaller 7202 Mayfair Creamer and the larger 7207 Beverly Creamer.

Four different creamers: Mayfair with ridged cherry wood handle (left); Beverly with straight top cherry wood handle (center); Beverly with ridged cherry wood handle (right); and original Mayfair with aluminum handle (top).

Comparison of the different Sugar Bowls.

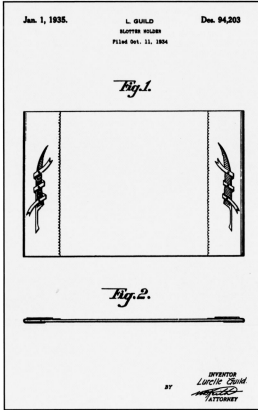

Jan. 1, 1935. L. GUILD Des. 94,203

BLOTTER HOLDER

Filed Oct. 11, 1934

Fig. 1.

Fig. 2.

INVENTOR
Lurelle Guild.

BY *ATTORNEY*

7000 Waverly Blotter Holder (retail $6.50), decorated with the thematic desk set emblem of brass quill and ribbon motif, on aluminum side frames with serrated edges, 12-1/2" x 20". $50-60

7006 Waverly Standard Size Blotter Holder (not pictured) is without serrated edges, 19" x 24," and described in the catalog: "Display this man's size blotter holder on a full-sized desk, and it lends a dignity that men like. Either the home library or the office will welcome this Kensington piece, designed to supplement the other Waverly desk appurtenances. Of course the felt protection on the under side prevents the possibility of harming the finest desk surface."

United States Patent Office drawing of Waverly Blotter Holder, designed by Guild, filed Oct. 11, 1934 and issued Jan. 1, 1935.

7001 Waverly Memorandum Book (retail $5.00), easily refillable, with brass quill and ribbon motif on front cover, with plain back, 6-1/2" x 8-1/2". $40-45
7005 Waverly Letter Opener (retail $1.00), aluminum quill, rare, 7-1/2". $35-40

United States Patent Office drawing for Waverly Memorandum Book, designed by Guild, filed Oct. 11, 1935 and issued Dec. 25, 1934.

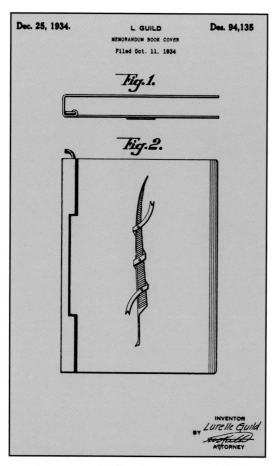

7002 Waverly Correspondence Rack (retail $4.75), with brass quill and ribbon riveted to the front, three slots, 7" x 4". $50-60

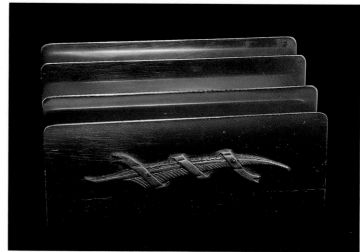

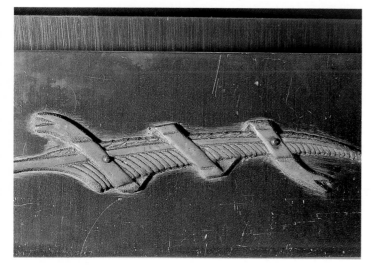

Detail of brass quill.

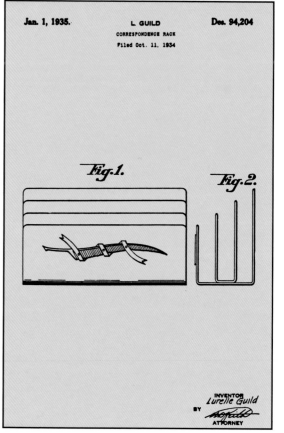

United States Patent Office drawing of Waverly Correspondence Rack, designed by Guild, filed Oct. 11, 1934 and issued Jan. 1, 1935.

United States Patent Office drawing of Waverly Ink &
Pen Holder, designed by Guild, filed Oct. 11, 1934 and
issued Dec. 25, 1934.

7003 Waverly Ink & Pen Holder, designed in 1934 (retail $5.00), with brass quill and
ribbon and square finial, with two horizontal grooves for pens, 4-3/4" x 4". $50-60

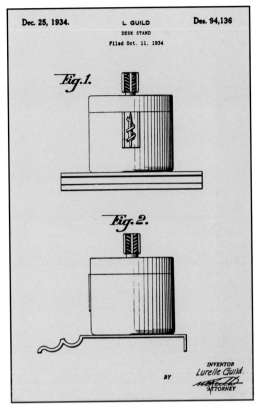

Detail of finial.

Detail of front emblem.

7004 Waverly Calendar Holder, with desk set emblem; this
example has quills facing the same direction, while the catalog
shows quills facing in opposite directions, 2-3/4" x 4-1/2". $25-35

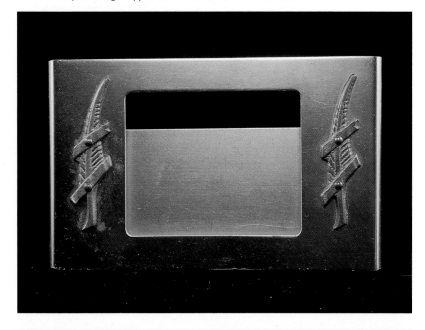

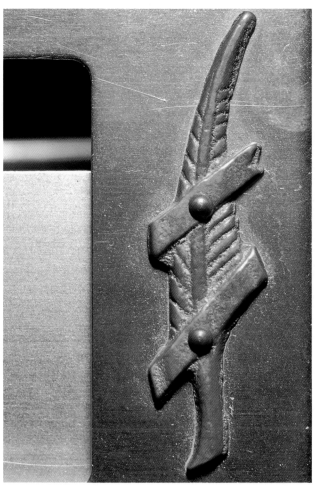

Detail of brass quill.

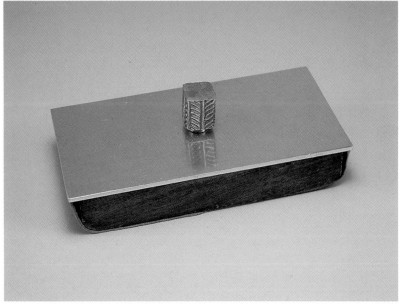

7008 Waverly Hand Blotter Holder (retail $2.25), square brass knob on
aluminum with wooden sides, 2-1/2" x 4-7/8". $30-35
(Not illustrated is 7009, a larger rocking version, 3-1/4" x 6-1/8".)

7007 Jot-It Memorandum Pad Holder (retail $1.00), holds 50-sheet pad, 4" x 7". $20-25

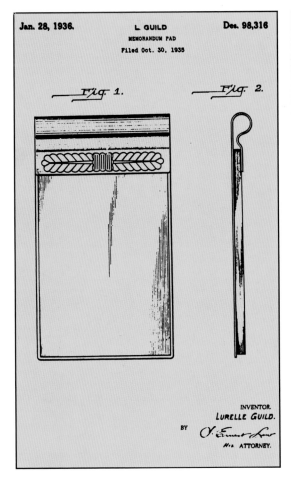

Detail.

United States Patent Office drawing of Jot-It Memorandum Pad Holder, designed by Guild, filed Oct. 30, 1935 and issued Jan. 28, 1936.

7012 Lido Lipstick Tissue Holder (retail $1.00), with female profile, described in the catalog: "To Milady's lips this trim little holder is dedicated. It neatly secures one package of Kleenex Lipstick Tissues, which are used when lipstick is applied, repaired, or renewed—with a great reduction in the casualties among handkerchiefs and towels. The tissues are also useful for keeping spectacles bright..." 2" x 3-1/4". $50-60

7013 Oxford Memo Roll (retail $5.00), described in catalog: "Here's an end to unsightly scribblings on phone book covers and little scraps of paper. The Oxford Memo Roll is a beautiful device which neatly provides almost inexhaustible material or telephone call notations, office memoranda, shopping lists, and the like. Its place is in the home (by the telephone) and in the office (on the busy executive's desk). It is made with carefully matched walnut side discs and polished brass guide strips. Standard adding-machine tape, replaceable at any stationery shop, is used." 9-1/4" l. $80-90

Waverly Ink & Pen Holder with Oxford Memo Roll.

United States Patent Office drawing of
Oxford Memo Roll, designed by Guild, filed
Nov. 2, 1939 and issued Feb. 13, 1940.

7440 Kenfold Money Minder (retail $2.50), issued as a billfold, fabric-lined and opens from each side, 3-3/4" x 3". $75-80

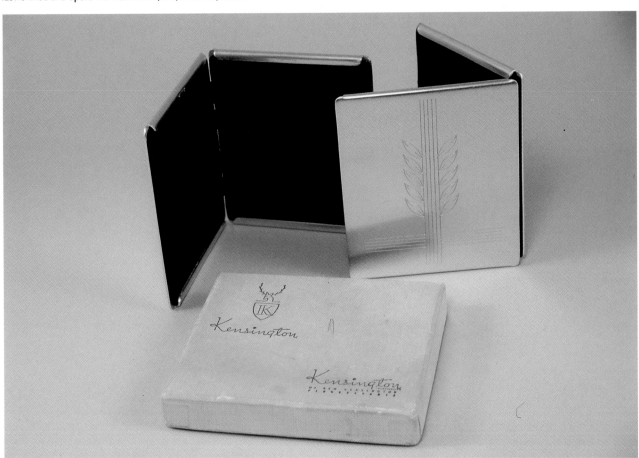

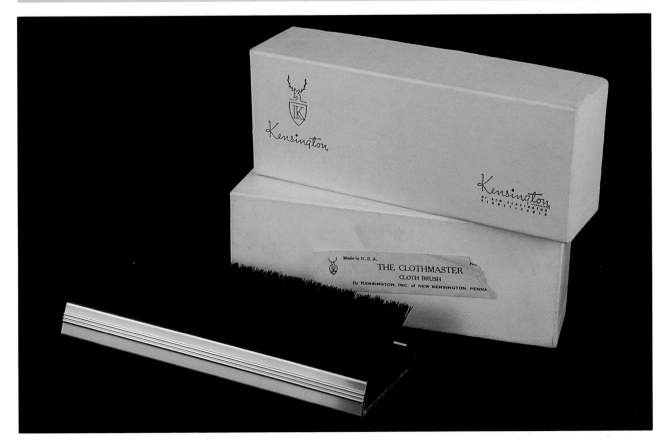

Opposite:

Wastebasket, unknown line number, 10" h. $80-90

Shown with 7252 Coldchester tumbler for size comparison, 5-1/4" h. $7-10

Above:

7693 Clothmaster Brush (retail $2.00), "...plays a big part in any household. Its bristles are strong and its size is ample to brush the coarsest or the finest fabric." 6-1/2" l. Shown with original box. (A shorter version, also with ribbed aluminum surface, called the Hatmaster, was also made.) $10-20

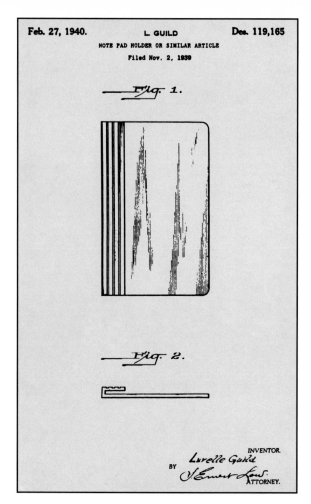

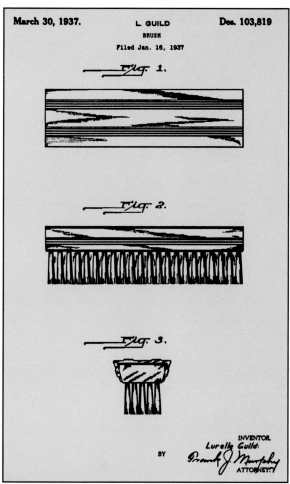

United States Patent Office drawing of a note pad holder, designed by Guild, filed Nov. 2, 1939 and issued Feb. 27, 1940; unknown line number or name.

United States Patent Office drawing of Clothmaster, designed by Guild, filed Jan. 16, 1937 and issued March 30, 1937.

7501 Rodney Memorandum Book, a plainer version of the Waverly, same size, 6-1/2" x 8-1/2". $30-35

7541 Hexagon Cigarette Box (retail $5.00), with round brass finial, holds thirty cigarettes, 4-3/4" h. $50-60

United States Patent Office drawing of Hexagon Cigarette Box, by Lurelle Guild, filed Oct. 11, 1934 and issued Jan. 8, 1935.

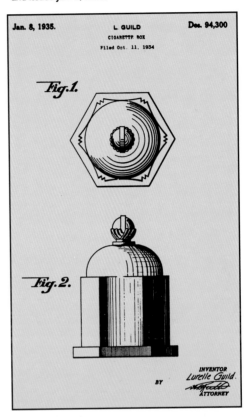

Detail of brass finial.

Original photograph of 7540 Dome Cigarette Box (retail $3.50), has brass finial, is smoothly finished, and holds a full pack of cigarettes, 4-3/4" h. *Photo courtesy Alcoa $50-60*

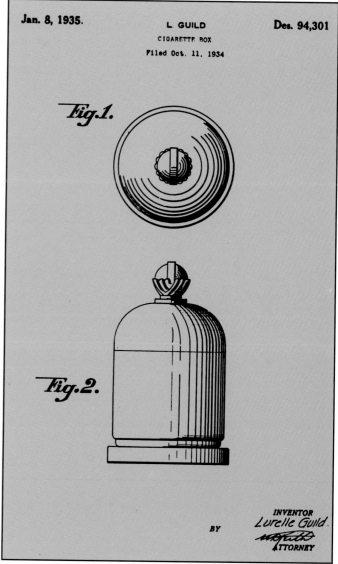

Jan. 8, 1935.

L. GUILD
CIGARETTE BOX
Filed Oct. 11, 1934

Des. 94,301

Fig.1.

Fig.2.

BY

INVENTOR
Lurelle Guild.

ATTORNEY

United States Patent Office drawing of Dome Cigarette Box, by Lurelle Guild, filed Oct. 11, 1934 and issued Jan. 8, 1935.

7610 Hexagon Ashtrays (retail four for $5.00), with brass doves and laurel branch mounted in center, 3" d. $15-20

United States Patent Office drawing of Hexagon Ash Tray, by Lurelle Guild, filed Oct. 11, 1934 and issued Jan. 8, 1935.

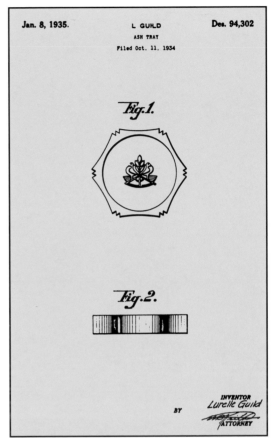

Detail of brass doves.

Original catalog photograph of 7471 Sussex Tobacco Jar (retail $5.00), with clip on underside of lid to hold a moist sponge, 4-3/4" h., 5-1/4" d. *Photo courtesy Alcoa* $40-50

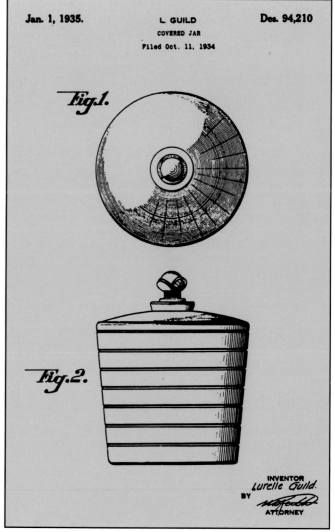

Jan. 1, 1935.   L. GUILD   Des. 94,210
COVERED JAR
Filed Oct. 11, 1934

Fig.1.

Fig.2.

INVENTOR
Lurelle Guild.
BY
ATTORNEY

United States Patent Office drawing of covered jar, by Lurelle Guild, filed Oct. 11, 1934 and issued Jan. 1, 1935; the finial design is slightly different, and the 1935 company publication suggests it is used for candy, nuts, crackers, and ice.

Variation of 7471 Sussex Tobacco Jar (retail $5.00), introduced in late 1939 with
clip on underside of lid to hold a moist sponge, 4-3/4" h., 5-1/4" d. $40-50
Shown with 7232 Riviera Pitcher.

Detail of finial attached directly to the lid, rather than on a raised ring as seen on the original design.

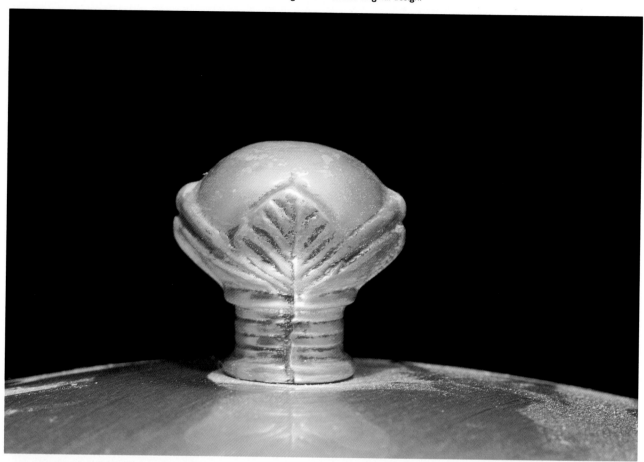

Another view of Sussex Tobacco Jar showing sponge clip in lid.

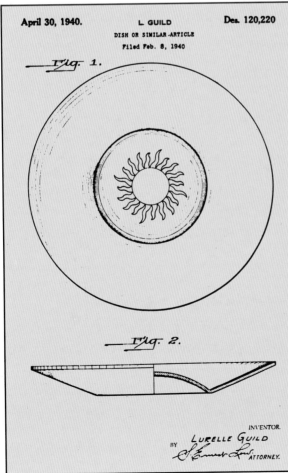

United States Patent Office drawing of Corona Ash Tray, by Lurelle Guild, filed Feb. 8, 1940 and issued April 30, 1940.

Above:

Left: 7617 Colby Ash Tray (retail $2.50); catalog description reads: "...safety first design. Cigarette rests on central pyramid. If left burning, it drops safely into the deep tray." 4-3/4" d. $15-20

Center: 7616 Empire Smoker's Tray (retail $2.50); catalog: "...really big enough to hold a pipe, and with a cork doodaddle in the center for knocking out the ashes...made of Kensington metal, they say it doesn't ever scuff, or get stained by burning tobacco...." 6-1/4" d. $20-30

Right: unknown line, appears to be part of a patriotic series with American eagle and shield surrounded by stars on a raised center, 4-5/8" d. $10-15

Front: 7617 Colby Ash Tray
Back: 7331 Shell Canapé Plate
Center: 7416 Corona Ash Tray (retail $3.50), with stylized
sun on raised center, also to serve candy, 7" d. $25-30

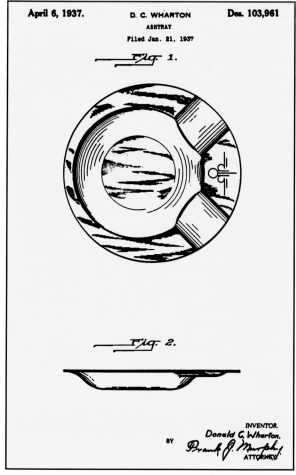

April 6, 1937.   D. C. WHARTON   Des. 103,961

ASHTRAY

Filed Jan. 21, 1937

Fig. 1.

Fig. 2.

INVENTOR.

Donald C. Wharton.

BY Frank J. Murphy

ATTORNEY.

Above:

Back: 7148 Chatham Tray

Front left: 7609 Mall Ashtray, designed by Donald G. Wharton (retail $1.00), with Art Deco motif, "with its off-center well and especially wide lip provides ample room for cigarette, cigar, or pipe. Burning tobacco does not stain Kensington Metal." 5" d. $5-10

Front center: 7609 Mall Ashtray without decorative motif. $5-10

Front right: 7609 Mall Ashtray with dolphin and trident, possibly from shipping line series. $5-10

United States Patent Office drawing of Mall Ashtray, designed by Donald G. Wharton, filed Jan. 21, 1937 and issued April 6, 1937.

7612 Royal Family Ashtrays (retail $1.50 each), fitted with cast snuffers in the form of King 7612-A, Queen 7612-B, Jack 7612-C, and Joker 7612-D. 4-1/4" d. $15-20 each

Detail of King.

Detail of Queen.

Detail of Joker.

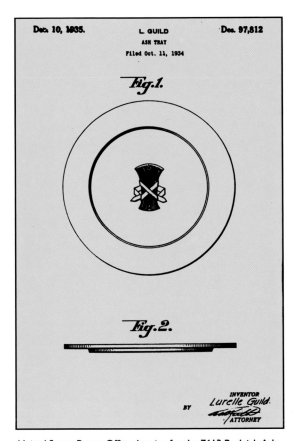

Dec. 10, 1935.

L. GUILD
ASH TRAY
Filed Oct. 11, 1934

Des. 97,812

Fig.1.

Fig.2.

INVENTOR
BY   Lurelle Guild.
ATTORNEY

United States Patent Office drawing for the 7613 Burleigh Ash Tray, with brass wheat sheaves wrapped in ribbon, filed Oct. 11, 1934 and issued Dec. 10 1935, but the item was discontinued after the first catalog, 5-1/2" d.

Back right: 7611 Ashtray with shipping line logo of gulls circling smoke stack, 5-3/8" d. $10-15
Front left: 7614-A Princess Anne Ashtray (retail $1.00), with wreath design, 4-3/8" d. $5-10
Back left: 7611-B Prince Edward Ashtray, designed by Du Pont Cornelius (retail $1.50), with classic medallion, 5-3/8" d. $10-15
Catalog description: These practical dished ash trays will retail their soft lustre right through countless snuffings of lighted cigarettes. For Kensington does not stain or tarnish. The permanence of this finish is a revelation. Either size comes in classic or wreath design. Larger size makes an excellent mint dish."

Detail of "classic medallion" motif.

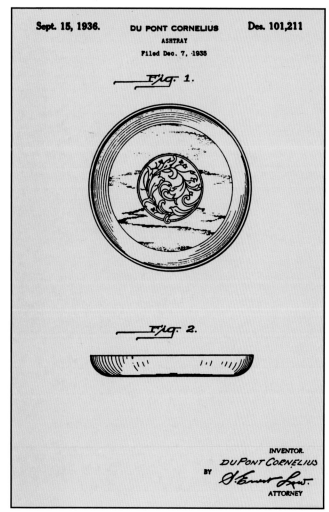

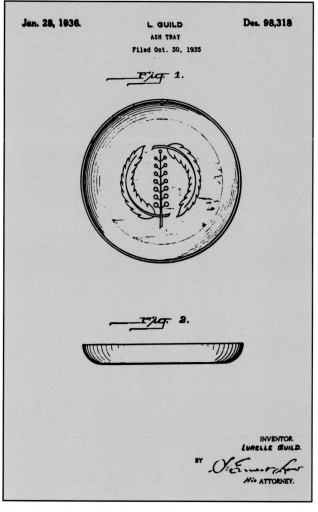

United States Patent Office drawing of Prince Edward Ashtray, designed by Du Pont Cornelius, filed Dec. 7, 1935 and issued Sept. 15, 1936.

United States Patent Office drawing of Princess Anne Ashtray, designed by Lurelle Guild, filed, filed Oct. 30, 1935 and issued Jan. 28, 1936.

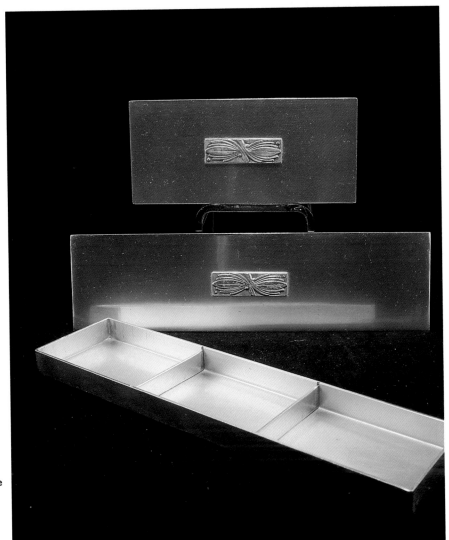

Top: 7547 Carolinian Cigarette Box (retail $7.50), with rectangular brass decoration mounted on the lid, can hold 60 loose cigarettes or three packs, 11-3/4" l. $50-60

Bottom: 7546 Virginian Cigarette Box, with rectangular brass decoration mounted on the lid, three compartments in bottom, 7-3/8" l. $50-60

Another view of the Virginian Cigarette Box.

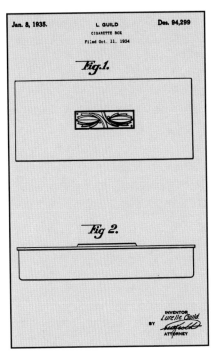

United States Patent Office drawing of Cigarette Box, by Lurelle Guild, filed Oct. 11, 1934 and issued Jan. 8, 1935.

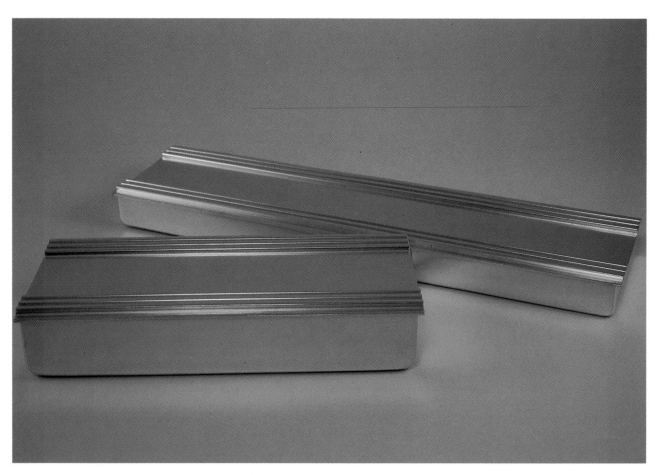

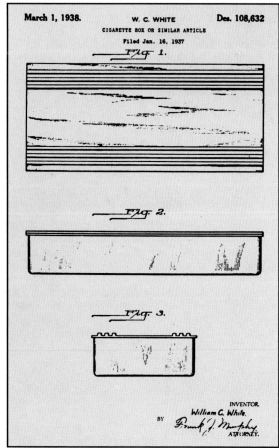

Above:
Bottom: 7548 Manor Cigarette Box, designed by William
C. White (retail $6.00), all metal with parallel ridges on
cover, holds three packs of cigarettes, 7-3/8" d. $40-50
Top: 7549 Town House Cigarette Box (retail $7.50), an
extended version of Manor, also used for pins and cards,
11-3/4" l. $45-55

United States Patent Office drawing of Cigarette
Box, by William C. White, filed Jan. 16, 1937 and
issued Mar. 1, 1938.

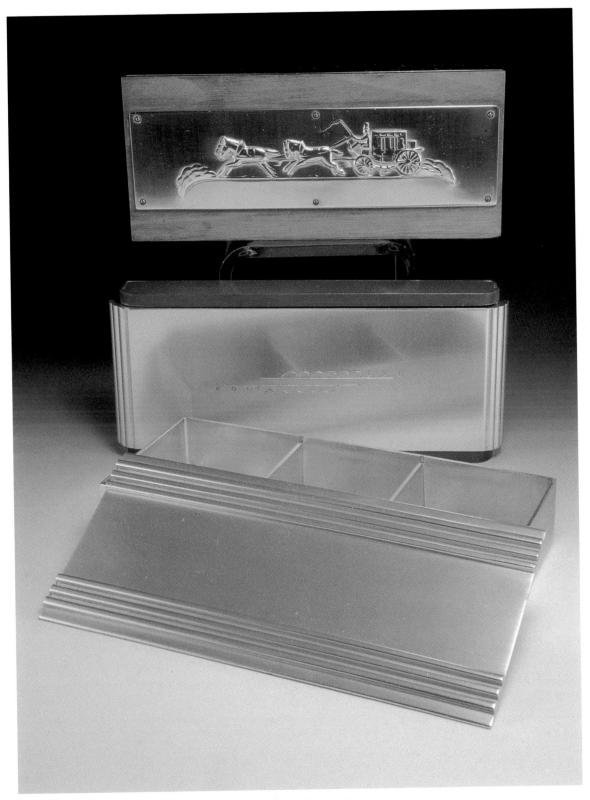

Top: 7547 Coach 'N Four Cigarette Box (retail $6.00), with repoussé design of stagecoach, top-hatted driver, and four horses galloping across a metal panel riveted to a walnut lid covering box with three interior compartments, 7-3/8" l. $45-55
Center: Buckingham Cigarette Box. Bottom: 7549 Town House Cigarette Box.

Detail of Coach 'N Four lid.

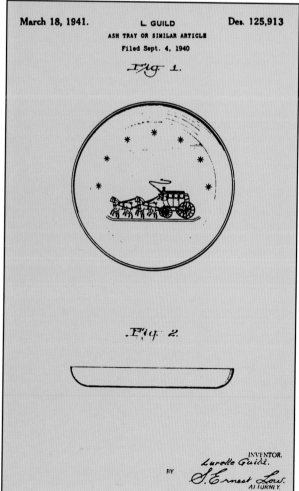

United States Patent Office drawing for 7610 Coach
'N Four Ashtray (retail $1.00), by Lurelle Guild, filed
Sept. 4, 1940 and issued March 18, 1941.

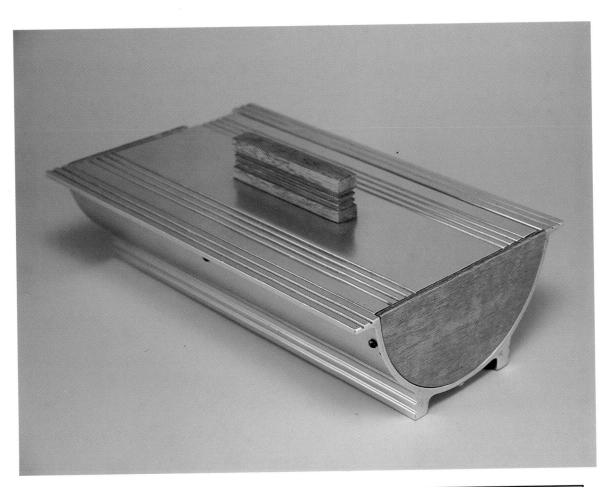

7542 Penthouse Cigarette Box (retail $7.50), walnut handle lifts ribbed aluminum lid of semi-circular base, with smooth metal sides, ribbed aluminum feet and top edges, and walnut ends, 6-1/2" l. $45-55

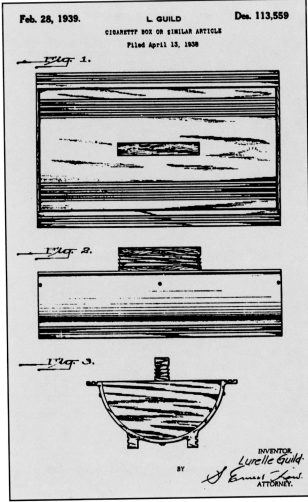

Feb. 28, 1939.   L. GUILD   Des. 113,559

CIGARETTE BOX OR SIMILAR ARTICLE
Filed April 13, 1938

Fig. 1.

Fig. 2.

Fig. 3.

INVENTOR.
Lurelle Guild.
BY
S. Ernest Jones
ATTORNEY.

United States Patent Office drawing of Penthouse Cigarette Box, by Lurelle Guild, filed April 13, 1938 and issued Feb. 28, 1939.

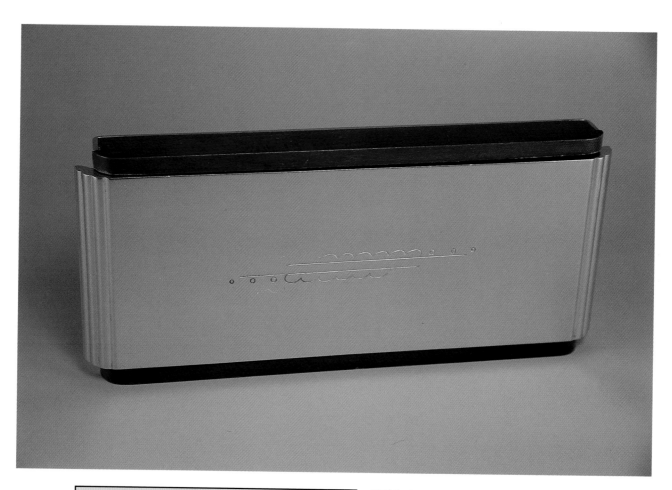

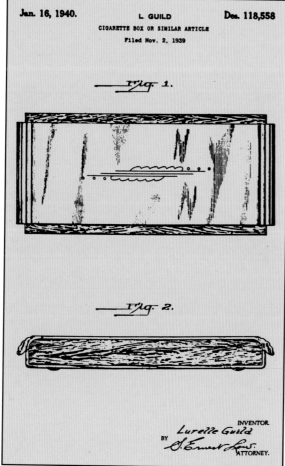

7546 Buckingham Cigarette Box (retail $5.00), the aluminum lid incised with circles and lines, with ridged edges, on metal and cherry wood base, holds 40 cigarettes, 7-3/8" l. $45-55

United States Patent Office drawing of Buckingham Cigarette Box, by Lurelle Guild, filed Nov. 2, 1939 and issued Jan. 16, 1940.

United States Patent Office drawing for Crest Match Box Holder, by Lurelle Guild, filed Oct. 11, 1934 and issued April 9, 1935.

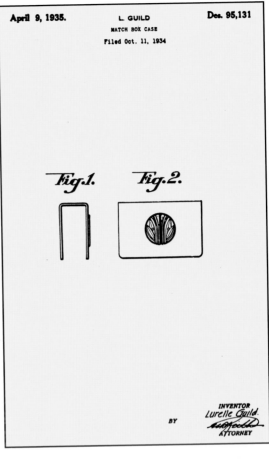

7605 Crest Match Box Holder (retail $1.00), with circular brass emblem, holds standard match box, 1-1/2" x 2-1/2". $10-15

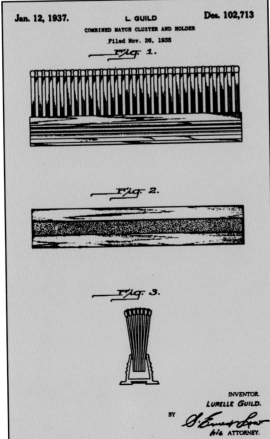

United States Patent Office drawing, for Rainbow Match Master, by Lurelle Guild, filed Nov. 26, 1935 and issued Jan. 12, 1937.

7620 Rainbow Match Master, represented in the fourth catalog as: "True to name, the match tips are arranged in rainbow-hued stripes. Concealed in a slot underneath is a replaceable strip of striking surface. Matchmaster is a decorative and useful novelty that has wide appeal. Doubly practical because refills are made available. Very popular as a bridge prize or favor. Although inexpensive, it is smart as only Kensington can be." 6-1/8" l. $15-20

7606 Vanity Fair Small Matchbox Holder (retail $0.50), plain and used at individual table setting or as a set on a tray with the Hostess Ashtray, 1-3/8" x 1-5/8"; 7607 Hostess Ashtray (retail $0.75), 2-5/8" d.; 7608 Hostess Set (retail $5.00), comprised of four each of Vanity Fair Small Matchbox Holders and Hostess Ashtrays in a lined box.

Opposite:
Smoking Stand, of aluminum with brass knob on top of handle for ash receiver, brass cross-hatching around top and bottom of pedestal; marked 5769 WEAREVER on bottom of ash receiver, 15-1/4" d., 26-1/2" h. Although the mark is Wearever, this is a typical Kensington design. $250-275

View of ash receiver with cover in lifted position; three-slot tray is 7-3/4" d.; three-inch ledge allows room for cocktail glasses, illustrating why a smoking stand of this design is sometimes referred to as a cocktail/smoker.

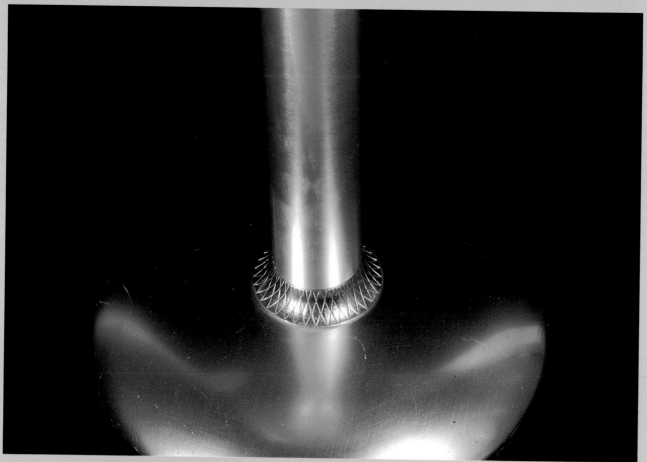

Detail of cross-hatching on brass trim at pedestal base.

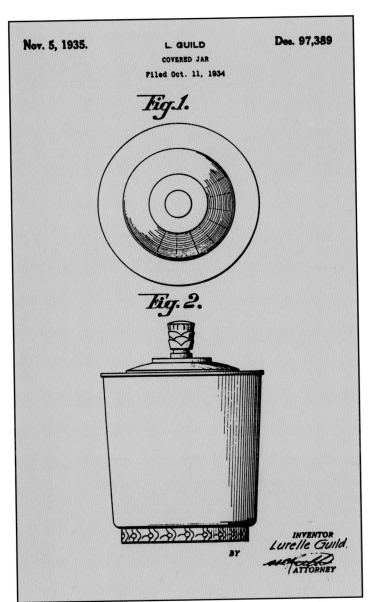

United States Patent Office drawing, for 7691 Charleston Tobacco Jar, by Lurelle Guild, filed Oct. 11, 1934 and issued Nov. 5, 1935, with brass knob on cover and band around base, (retail $6.00) 5" h.

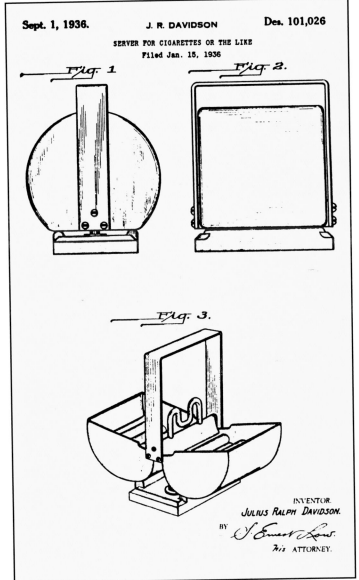

United States Patent Office drawing, for 7550 Piccadilly Cigarette Server, designed by Julius Ralph Davidson, a German citizen residing in Chicago, filed Jan. 15, 1936 and issued Sept. 1, 1936. The catalog description is as follows: "The Piccadilly is decidedly clever. A lift of the handle, and the cylinder opens up and at you, so to speak, like a clamshell, each half holding a full package of cigarettes. Push down, and The Piccadilly is closed. Nothing to get out of order. Original, beautiful, and surprising in its cleverness. Contrasting black Bakelite base." (retail $5.00), 2-pack capacity.

Furniture

Kensington made aluminum furniture for only five years, from 1947 to 1952.

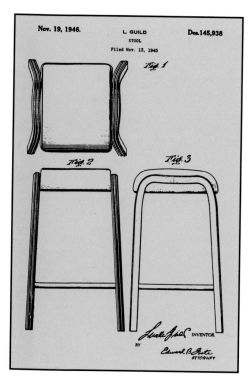

United States Patent Office drawing of a Stool, designed by Lurelle Guild, filed Nov. 13, 1945 and issued Nov. 19, 1946, with Kensington aluminum frame and upholstered seat.

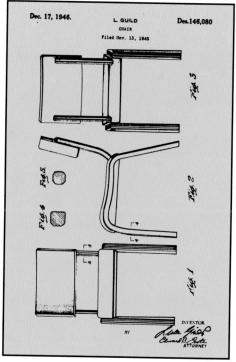

United States Patent Office drawing of a Chair, designed by Lurelle Guild, filed Nov. 13, 1945 and issued Nov. 17, 1946, with Kensington aluminum frame and upholstered seat and back.

United States Patent Office drawing of an aluminum Cabinet, designed by Lurelle Guild and Carlton G. Towne, filed July 26, 1947 and issued April 23, 1949.

Moiré

When the novelty of machine-age, Art Deco designs began to fade, and the expense of manufacturing cast brass accents became impractical, Kensington introduced a line of giftware called Moiré. By using simple mid-century styling, including modern freeform shapes, the "etched" moiré pattern gave Kensington metal a few more years of shelf life. (Kensington was identified as a division of Wear-Ever Aluminum, Inc. in the 1960 Moiré brochure.) The new product was complete with the usual promotional promises in company publications:

> "Kensington Moiré Giftware, the metal of lifelong beauty...Kensington introduces a new, popularly priced giftware line in distinctive Moiré. This unique, crystal-like etched appearance is produced on a special aluminum alloy. The lustrous sheen in silver or gold is achieved through alumiliting. This process also gives the metal extra hardness, insuring long life and greater resistance to smudges, stains, and normal scuffing. Kensington Moiré is solid metal, intrinsically lovely, with no lacquer coating or plating to chip or peel. Minimum care is required to maintain its permanent, natural beauty. Simply wash Kensington Moiré in mild sudsy water, rinse and dry."

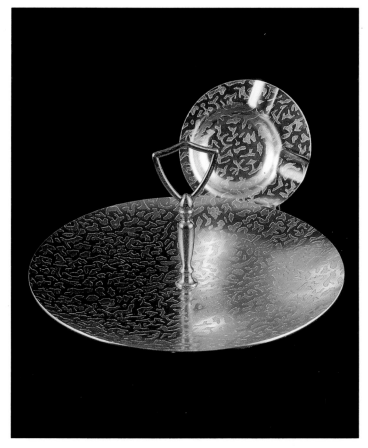

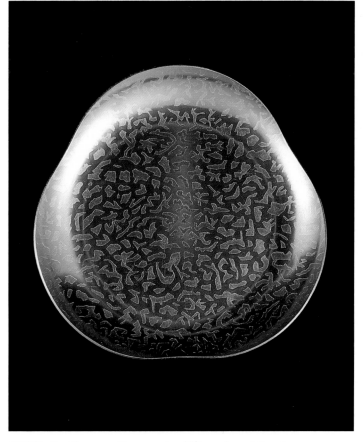

Back: G-7780 Westbrook Ash Tray, "popular convenient size," which was really the Mall Ashtray with a new name, 5"d. $5-10
Front: G-7741 Hastings Plate, "smartly styled, gently sloped serving plate with center fluted stem and handle...for petits fours, canapes or party sandwiches." 10" d. $15-20

7749 Stratford Server, multi-purpose bowl "for candies, mints, nuts, snacks...bridal shower gift or bridge prize." 9-1/4" d. $20-30

Detail of Silver Moiré pattern.

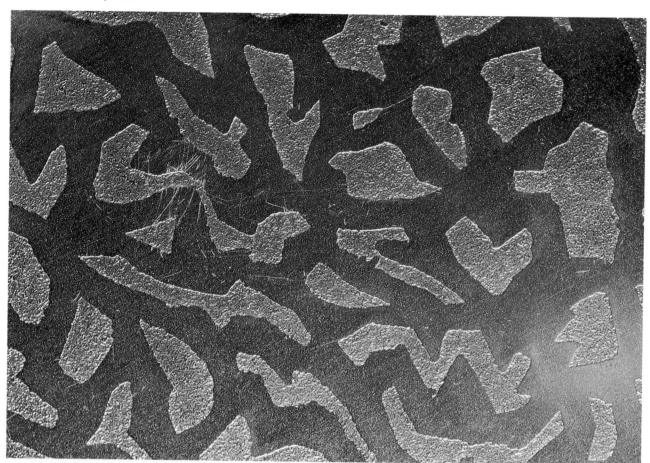

7754 Townsend Compote, smooth cone-shaped base
supports freeform bowl, 9-1/4" d., 4-1/2" h. $25-35

G-7752 Georgian Tiered Tray, "superbly styled in modern, free form design to enhance any table setting," the base is another use for the Stratford Server, 9-1/4" d., 10-1/2" h. $30-40

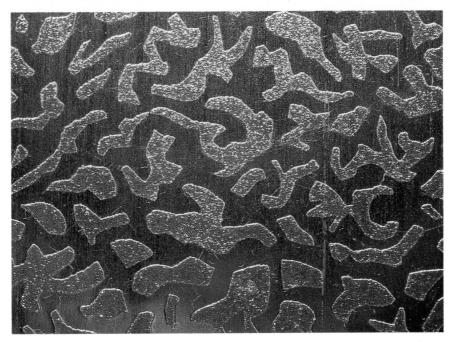

Detail of Gold Moiré pattern.

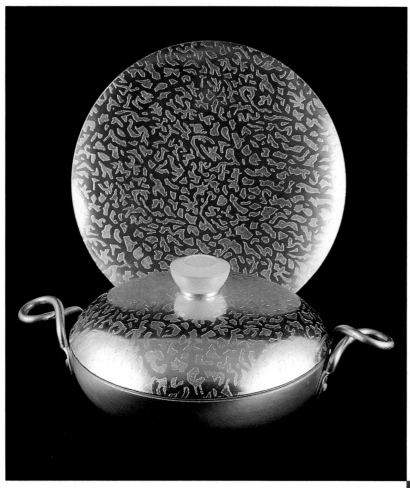

Back: 7740 Norwich Canape, for canapes, petits fours, relishes, or as a wall or mantle decoration, 10" d. $15-20

Front: Covered serving bowl with smooth aluminum base with twisted handles and Silver Moiré lid with ivory Bakelite knob, 8" d. (12" d. with handles). $25-35

7751 Carlton Server, 9-1/4" d. (also available as 7750 in 8" d.) $10-15

Alcoa was not the only company to succumb to the pressure from a significant segment of the American public, that either wanted to return to the security of traditional forms or had never left it in the first place. While it is certainly possible that a percentage of the market for original Art Deco Kensington later tried the "fashionable" Moiré, it would be surprising to find the same consumers reverting back to the "new" traditional line. A more likely (and undeniably large) market for these unoriginal items would be those who furnished with colonial revival rather than, say, Herman Miller or Knoll.

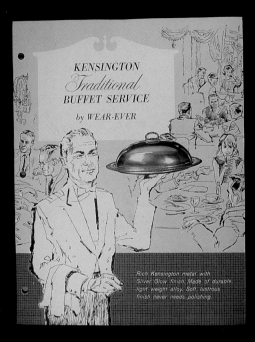

Catalog pages for Kensington Traditional
Buffet Service by Wear-Ever

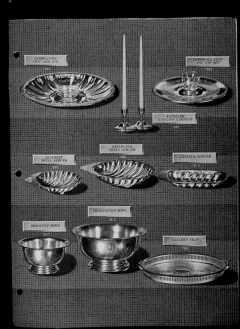

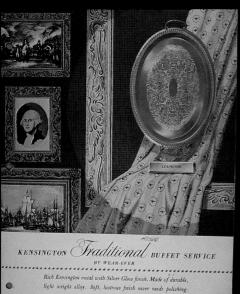

KENSINGTON *Traditional* BUFFET SERVICE
BY WEAR-EVER

Rich Kensington metal with Silver Glow finish. Made of durable, light weight alloy. Soft, lustrous finish never needs polishing.

LEXINGTON Cat. No. 7840. This large oval tray fitted with handles and measuring 17" x 22¼", serves for standing rib as well as it will for sandwiches and hors d'oeuvres.

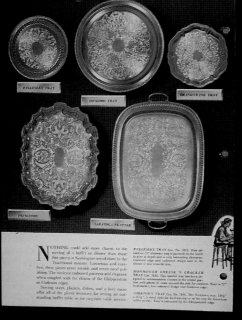

NOTHING could add more charm to the serving of a buffet or dinner than these fine pieces in Kensington metal done in the Traditional manner. Luxurious and care-free, these pieces never tarnish and never need polishing. The intricate embossed patterns add elegance when coupled with the charm of the Chippendale or Gadroon edges.

Serving trays, platters, dishes, and a lazy susan offer all of the pieces necessary for setting an outstanding buffet table or for exquisite table service.

WELLESLEY TRAY Cat. No. 7825. This delicately grooved 13" diameter tray is grooved on the inside to give it depth and a very interesting character. Gadroon edge and embossed design add to the charm of this versatile tray.

MONMOUTH CHEESE 'N CRACKER TRAY Cat. No. 7826. This useful tray has been designed to accommodate cheeses in the center position with plenty of room around the side for crackers. Tray is 12" in diameter, has embossed design and Gadroon edge.

YORKTOWN TRAY Cat. No. 7831. The Yorktown tray, 13¼" x 21¼", is sized right for hold serving or as the tray for luxurious coffee service. Tray is embossed by the Chippendale edge.

Catalog pages for Kensington Traditional
Buffet Service by Wear-Ever

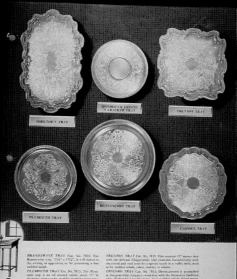

BRANDYWINE TRAY Cat. No. 7835. The Brandywine tray, 12¼" x 17¼", is well suited to the serving of appetizers, or for presenting a fine molded stand.

PLYMOUTH TRAY Cat. No. 7823. The Plymouth tray is set all around utility piece; 13" in diameter, that can be used for anything from serving drinks to displaying a cake. Has embossed design and graceful Gadroon edge.

SARATOGA PLATTER Cat. No. 7840. Massive 19" x 29", this platter is ideally suited to the serving of fowl, roasts, cold meats, and seafoods. Fitted with decorative handles.

PRINCETON Cat. No. 7839. For complete elegance, this large oval tray with the Chippendale edge is unmatched. Size is 17¾" x 21½".

TRENTON TRAY Cat. No. 7830. This unusual 13" square tray with the delicate Chippendale edge contrasts harmoniously with the round and oval trays for a special touch at a buffet table, such as for molded salads, cakes, cookies, or snacks.

CONCORD TRAY Cat. No. 7825. Dimensionenent is permitted in this gracefully designed round tray with the decorative Gadroon edge. Perfect for hors d'oeuvres, large molded salads, sliced meats. Diameter 18½".

KENSINGTON TRAY Cat. No. 7822. This 15" diameter tray will serve appetizers, pastries, and sweets with an air of good taste. Delicate embossed design is supplemented with decorative, Gadroon edge.

CAMDEN TRAY Cat. No. 7836. The Chippendale edge is blended tastefully with the refined embossed pattern in this unusual octagonal tray. Overall dimensions are 14⅝" x 14¼".

#7846
–Silver Glow
#G7846
–Golden Glow

Lexington Tray. This large oval tray, fitted with handles, serves for standing rib as well as it will for sandwiches and hors d'oeuvres.

Size: 17" x 22¼"
Unit Pack: 1
Carton Weight: 4.7 lbs.

#7878
–Silver Glow
#G7878
–Golden Glow

Cornell Bread Tray. Generously proportioned to hold bread, toast, rolls, or muffins.

Size: 12¼" x 6¾" x 1¼" deep
Unit Pack: 1
Carton Weight: 1.0 lbs.

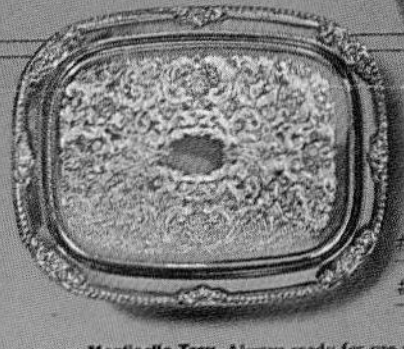

#7883
–Silver Glow
#G7883
–Golden Glow

Monticello Tray. Always ready for use when occasions call for canapes, pastries, or beverages.

Size: 14" x 18" x ½" **Unit Pack:** 1 **Carton Weight:** 2.2 lbs.

#7890
–Silver Glow only

Cornwall Chip and Dip Set. A popular serving piece with Traditional charm. Center glass bowl holds favorite dip for encircling array of crackers, chips and pretzels.

Size: 13" diameter x ¾" deep
Unit Pack: 1
Carton Weight: 2.0 lbs.

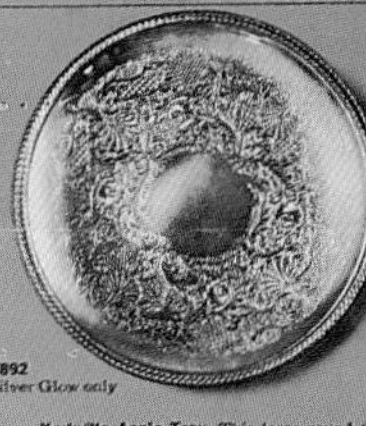

#7892
–Silver Glow only

Yorkville Aspic Tray. This large round tray is ideal for serving molds of meat, fish, or tomato aspic. Always ready for use, it never needs polishing.

Size: 18-inch diameter **Unit Pack:** 1 **Carton Weight:** 3.0 lbs.

#7891
–Silver Glow only

Gallery Tray. Ideally suited for the serving of all types of drinks, this tray lends a distinctive air of luxurious elegance for dining pleasure.

Size: 15½" diameter x 1⅜" deep
Unit Pack: 1
Carton Weight: 2.2 lbs.

WEAR·EVER SUBSIDIARY OF ALCOA
CHILLICOTHE, OHIO 45601

111-179

#150D

Wear-Ever Traditional

all the elegance you'll ever need for the buffet

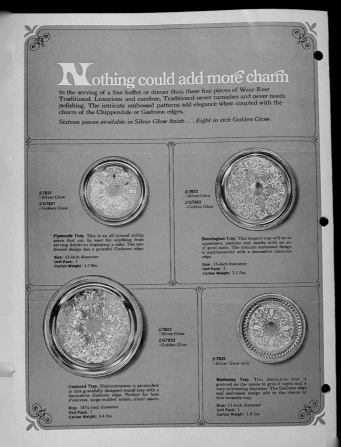

Nothing could add more charm

to the serving of a fine buffet or dinner than these fine pieces of Wear-Ever Traditional. Luxurious and carefree, Traditional never tarnishes and never needs polishing. The intricate embossed patterns add elegance when coupled with the charm of the Chippendale or Gadroon edges.

Sixteen pieces available in Silver Glow finish . . . Eight in rich Golden Glow.

#7821
–Silver Glow
#G7821
–Golden Glow

Plymouth Tray. This is an all around utility piece that can be used for anything from serving drinks to displaying a cake. The embossed design has a graceful Gadroon edge.

Size: 13-inch diameter
Unit Pack: 1
Carton Weight: 1.7 lbs.

#7822
–Silver Glow
#G7822
–Golden Glow

Bennington Tray. This elegant tray will serve appetizers, pastries and snacks with an air of good taste. The delicate embossed design is supplemented with a decorative Gadroon edge.

Size: 15-inch diameter
Unit Pack: 1
Carton Weight: 2.1 lbs.

#7823
–Silver Glow
#G7823
–Golden Glow

Concord Tray. Distinctiveness is personified in this gracefully designed round tray with a decorative Gadroon edge. Perfect for hors d'oeuvres, large molded salads, sliced meats.

Size: 18¾-inch diameter
Unit Pack: 1
Carton Weight: 3.4 lbs.

#7825
–Silver Glow only

Wellesley Tray. This distinctive tray is grooved on the inside to give it depth and a very interesting character. The Gadroon edge and embossed design add to the charm of this versatile tray.

Size: 13-inch diameter
Unit Pack: 1
Carton Weight: 1.8 lbs.

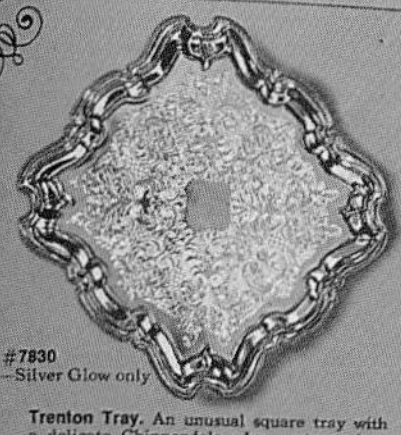

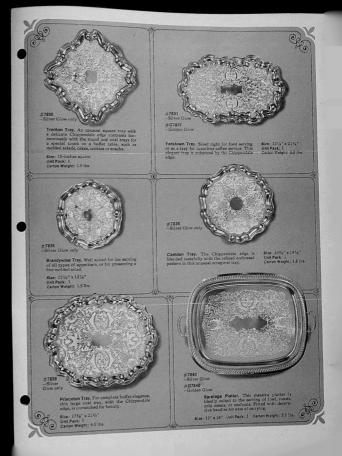

#7830
–Silver Glow only

Trenton Tray. An unusual square tray with a delicate Chippendale edge contrasts harmoniously with the round and oval trays for a special touch on a buffet table, such as molded salads, cakes, cookies or snacks.

Size: 15-inches square
Unit Pack: 1
Carton Weight: 1.6 lbs.

#7831
–Silver Glow
#G7831
–Golden Glow

Yorktown Tray. Sized right for food serving or as a tray for luxurious coffee service. This elegant tray is enhanced by the Chippendale edge.

Size: 13¼" x 21¼"
Unit Pack: 1
Carton Weight: 3.3 lbs.

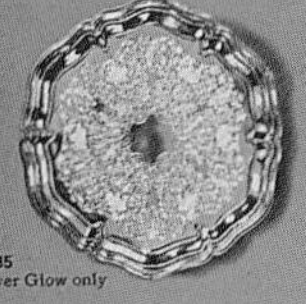

#7835
–Silver Glow only

Brandywine Tray. Well suited for the serving of all types of appetizers, or for presenting a fine molded salad.

Size: 12½" x 12⅞"
Unit Pack: 1
Carton Weight: 1.5 lbs.

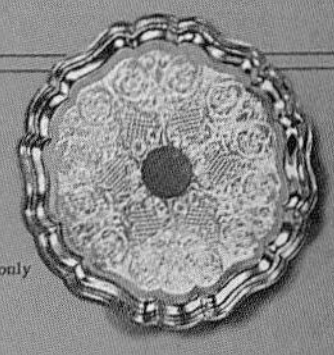

#7836
–Silver Glow only

Camden Tray. The Chippendale edge is blended tastefully with the refined embossed pattern in this unusual octagonal tray.

Size: 14⅝" x 14⅞"
Unit Pack: 1
Carton Weight: 1.8 lbs.

#7839
–Silver Glow only

Princeton Tray. For complete buffet elegance, this large oval tray, with the Chippendale edge, is unmatched for beauty.

Size: 17⅜" x 21¾"
Unit Pack: 1
Carton Weight: 4.0 lbs.

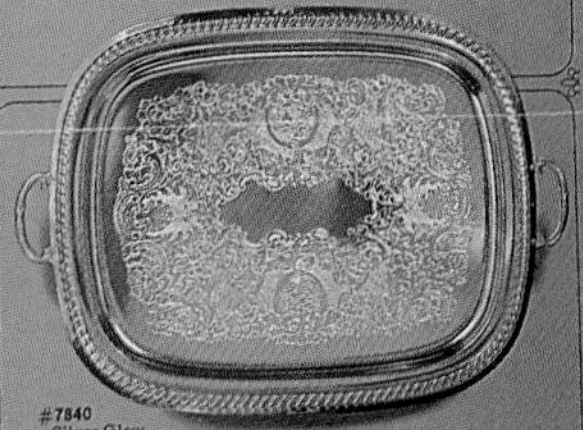

#7840
–Silver Glow
#G7840
–Golden Glow

Saratoga Platter. This massive platter is ideally suited to the serving of fowl, roasts, cold meats, or seafoods. Fitted with decorative handles for ease of carrying.

Size: 19" x 24" **Unit Pack:** 1 **Carton Weight:** 5.3 lbs.

Brandywine Tray, a "silver" look-alike with scalloped rim and historic
decoration, part of the Kensington "Traditional" service line, 12-3/4" d. $20-25

Detail of surface decoration.

Wear-Ever also produced a line called Buffet Service by Wear-Ever. A 1957 brochure pictures items in both familiar Guild-Kensington shapes and bland modified versions. Art Deco has been left in the past.

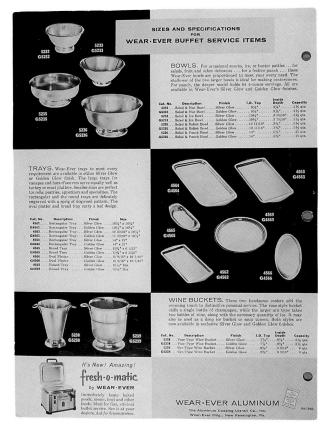

1957 catalog pages for Buffet Service by Wear-Ever.

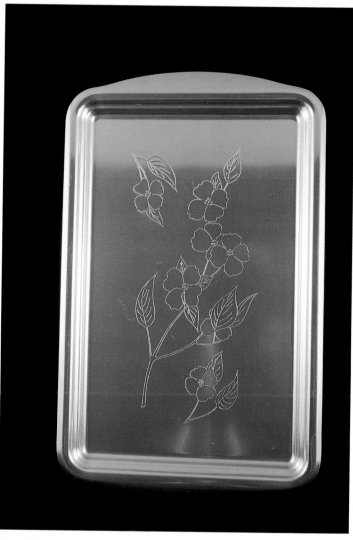

Wear-Ever Buffet Service 5238 Wine Bucket with twisted handles, similar to those used on some Moiré items, and a pedestal foot, 10" h. $30-35

Wear-Ever Buffet Service G-4563 Gold Rectangular Tray, with floral decoration, 11-3/4" x 18-3/4". $15-20

Selected Bibliography

Books/ Exhibition catalogs:

Campbell, Bonita. *Depression Silver: Machine Age Craft and Design.* exhibition catalog. Northridge: California State U., 1995.

Carr, Charles C. *Alcoa: An American Enterprise.* New York: Rinehart & Co., 1952.

Cheney, Sheldon and Martha Chandler. *Art and the Machine: An Account of Industrial Design in 20th-century America.* New York: Whittlesey House, 1936.

Greif, Martin. *Depression Modern: The Thirties Style in America.* New York: Universe, 1975.

Grist, Everett. *Collectible Aluminum: An Identification and Value Guide.* Paducah, Kentucky: Collector Books, 1994.

Kilbride, Richard J. *Art Deco Chrome: Book II.* Stamford, Connecticut: JO-D Books, 1992.

Meikle, Jeffrey. *Twentieth-Century Limited.* Philadelphia: Temple U. Press, 1979.

Museum of Modern Art. *Machine Art.* exhibition catalog. New York: Museum of Modern Art, 1934.

Pulos, Arthur J. *American Design Ethic: A History of Industrial Design to 1940.* Cambridge: MIT Press, 1983.

———-. *The American Design Adventure 1940-1975.* Cambridge: MIT Press, 1988.

Smith, George David. *From Monopoly to Competition: The Transformation of Alcoa 1888-1986.* Cambridge: Cambridge University Press, 1988.

Wilson, Richard Guy, et al. *The Machine Age in America 1918-1941.* New York: Brooklyn Museum and Harry N. Abrams, 1986.

Woodard, Dannie A. *Hammered Aluminum Hand Wrought Collectibles, Book Two.* Wolfe City, Texas: Hennington Pub., 1993.

Articles/sources of articles:

Alcoa. *The Alcoa News.* Oct. 15, 1934; Nov. 12, 1934; Nov. 11, 1935; Oct. 5, 1936; May 17, 1948; Sept. 5, 1949.

The Aluminist. newsletter, Box 1346, Weatherford, Texas 76086.

China & Glass. monthly trade journal. 1941, 1942 issues.

DeAngelo, Connie. "Decorative Aluminum." *The National Journal.* (Feb. 1981): 16.

"Decorative Aluminum." *The National Glass, Pottery and Collectables Journal.* (Sept. 1980): 37.

"Designer for Mass Production." *Art & Industry.* (June, 1938): 228-33.

Guild, Lurelle V.A. "Tableware Goes Zodiac." *Modern Home* 7 (April 1935):2.

Johnson, Frances. "Aluminum Ware—Hammered or Polished." *The Antique Trader Weekly.* (Jan 22, 1992):81-83.

———-"Aluminum." *Antiques & Collecting.* (Jan. 1, 1995):28-30, 49.

Monarch, Jerry. "Monarch's Miscellany." *Depression Glass Daze.* (Jan. 1981): 32-33.

Nelson, George. "Both Fish and Fowl." *Fortune.* 9 (Feb. 1934): 40-43, 80, 90-98.

Smith, Gregory W. "Alcoa's Aluminum Furniture: New Application for a Modern Material 1924-1934." *Pittsburgh History* 78 (Summer 1995): 52-64.

Wickware, Francis Sill. "Durable Goods Go to Town!" *Forbes* 37 (Nov. 15, 1936): 32-36, 70.

Archives and Primary Sources:

Alcoa Archives. Alcoa. Pittsburgh: miscellaneous correspondence, photographs, brochures, clippings, etc. and all volumes of *The Alcoa News.* The majority of the material was deposited at the Historical Society of Western Pennsylvania.

"Chairs by Kensington, Inc." typescript dated Aug. 22, 1946.

Guild, Lurelle. Letter to Anna Lydon of Alcoa, dated Sept. 28, 1979.

Kensington Ware company catalogs. 1930s and 1940s.

Lurelle Guild Archives. Special Collections, the Bird Library, Syracuse University: 92 boxes, 121 linear feet of material dating from 1931 to 1968.

Trump, H. S. (of Kensington, Inc.). Letter to Lurelle Guild, dated Jan. 20, 1938.

United States Patent Office Records.

"Wear-Ever Aluminum, Inc. History." typescript. Wear-Ever Aluminum, Chillicothe, Ohio, [1979].

Appendix A: Kensington Ware Items

Numbers and names are taken from company publications, such as catalogs and supplements. Occasionally, names were changed or a number was used for two items without explanation. With further research, these questions may be answered, and additional items may be found.

7000	Waverly Blotter Holder—12-1/2" x 20"
7001	Waverly Memorandum Book—6-1/2" x 8-1/2"
7002	Waverly Correspondence Rack—4 x 7"
7003	Waverly Ink and Pen Holder—4 x 4-3/4"
7004	Waverly Calendar—2-3/4" x 4-1/2"
7005	Waverly Letter Opener—7-1/2" l.
7006	Waverly Standard Size Blotter Holder—19 x 24"
7007	Jot-It Memorandum Pad Holder—4 x 7"
7007 1/2	Memo Refill
7008	Waverly Hand Blotter—2-1/2" x 4-7/8"
7009	Waverly Hand Blotter—3-1/4" x 6-1/2"
7010	Jot-It Pocket Memo Book—2-1/4" x 4-1/4"
7010 1/2	Memo Refills
7011	Shopper Note Book
7012	Lido Lipstick Tissue Holder—2 x 3-1/4"
7013	Oxford Memo Roll—9-1/4" l.
7014	Tab Desk Memo
7027	Laurel Vase—7"
7028	Laurel Vase—9"
7029	Gainsborough Vase—9" h.
7030	Sherwood Vase—9-1/2" h.
7031	Marlborough Vase—10" h.
7032	Kingston Flower Vase—6-1/4" h.
7050	Golden Chatham Tray—8-1/2" x 14"
7060	Golden Dorchester Tray—10-3/4" sq.
7100	Zodiac platter—18" d.
7102	Compass Platter—15" d.
7103	Yorkshire Covered Cake Tray—15-3/8" d.
7104	York Cheese and Cracker Server—15-3/8" d.
7105	Aztec Footed Platter—16" d.
7106	Cortez Footed Platter—12" d.
7107	Constellation Platter—18" d.
7108	Laurel Tray—12" d.
7109	Laurel Tray—15" d.
7110	Laurel Tray—18" d.
7111	Mayfair Round Tray—12" d.
7112	Mayfair Round Tray—15" d.
7113	Tray
7148	Chatham Tray—8-1/2" x 14"
7149	Coach'n Four Tray—6 x 11"
7150	Radcliffe Tray—6-7/8" x 9-3/4"
7151	Chelsea Serving Tray—10-1/2" x 18"
7151 1/2	Chelsea Buffet Server—10-1/2" x 18"
7152	Clipper Ship Tray—14-3/8" x 22-5/8"
7152 1/2	Clipper Ship Buffet Server—
7153	Clipper Ship Tray—10-1/2" x 18"
7153 1/2	Clipper Ship Buffet Server—10-1/2" x 18"
7154	Hunt Serving Tray—11-3/4" x 20-1/2"
7155	Dover Bread Tray—13-1/4" x 6-1/8"
7156	Hunt Serving Tray—14-3/8" x 22-5/8"
7156 1/2	Hunt Buffet Server
7157	Chelsea Serving Tray—11-3/4" x 20-1/2"
7158	Irvington Bread Tray—6-1/8" x 13"
7159	Savoy Round Tray—12-1/4"
7160	Engine Tray (Pioneer)—10-3/4" sq.
7161	Tray
7162	Dorchester Tray—10-3/4" sq.
7163	Savoy Square Tray—10-3/4" sq.
7164	Commodore Perry Tray—10-3/4" sq.
7165	Queen Tray—10-3/4" sq.
7166	Square Tray, Plain—10-3/4" sq.
7200	Mayfair Coffee Server—12 cup, 10" h.
7201	Mayfair Tea Server—6 cup, 7" h.
7202	Mayfair Creamer—3-1/8" h.
7203	Mayfair Sugar—3-1/2" h.
7204	Sugarac Sugar Service
7205	Crackerac Cracker Service
7207	Beverly Creamer—3-1/4" h.
7231	Mayfair Water Pitcher—2 qt.
7231-P	Water Pitcher Plain—2 qt.
7232	Riviera Pitcher—2 qt. 7-1/2" h.
7244	Bowl
7245	Newport Ice Bowl—6" d.
7245	Coldchester Ice Tongs
7246	Coldchester Ice Bowl—7-7/8" d.
7247	Coldchester Ice Tongs—6-1/8" l.
7249	Coldchester Old-Fashioned Cup—3" d.
7250	Coldchester Wine Cooler—10-1/4" h.
7251	Coldchester Cocktail Shaker—1 1/2 qt.
7252	Coldchester Tumbler—14 oz. 5-1/4" h.
7253	Coldchester Cocktail Cup
7254	Five O'Clock Canape Server 6" d.
7255	Riviera Tumblers—14 oz.
7259	Epicurean Salad Bowl—11-5/8"
7260	Round Table Punch Bowl—12-1/4" d.
7261	Southampton Sauce Boat—7" l.
7300	Vanity Fair Coaster Plain
7300	Vanity Fair Coaster Decorated
7301	Vanity Fair Butter Plate—5-1/2" d.
7314	Clifton Oval Tray—11-1/4" l.
7315	Plain Sandwich Plate—10" d.
7315-D	Zodiac Sandwich Plates—10" d.
7316	Nassau Canape Plate—10" d.
7316-P	Plain Service Plate—10" d.
7317	Plain Service Plate—11" d.
7317-A	Zodiac Service Plate, Aquarius—11" d.
7317-B	Zodiac Service Plate, Pisces—11" d.
7317-C	Zodiac Service Plate, Aries—11" d.
7317-D	Zodiac Service Plate, Taurus—11" d.

7317-E	Zodiac Service Plate, Gemini—11" d.	7427	Ship's Galley Bowl—9-1/4" d.
7317-F	Zodiac Service Plate, Cancer—11" d.	7428	Imperial Fruit Bowl—10" d.
7317-G	Zodiac Service Plate, Leo—11" d.	7429	Holiday Bowl—6" d.
7317-H	Zodiac Service Plate, Virgo—11" d.	7430	Continental Covered Bowl—7" d.
7317-I	Zodiac Service Plate, Libra—11" d.	7431	Golden Ming Bowl—7" d.
7317-J	Zodiac Service Plate, Scorpio—11" d.	7432	Golden Dorchester Bowl—9-1/4" d.
7317-K	Zodiac Service Plate, Sagittarius—11" d.	7440	Kenfold Money-Minder—3 x 3-3/4"
7317-L	Zodiac Service Plate, Capricorn—11" d.	7470	Snack Cracker Jar
7318	Marquee Plate Cover—6-7/8" d.	7471	Sussex Candy Jar—4-3/4" h. 5-1/4" d.
7318	Marquee Hot Service Center	7472	Nottingham Jar—5-1/2" h. 5" d.
7320	Northumberland Canape Plate (Stag)—10" d.	7490	Thistleton Covered Bowl—4-3/4" h.
7321	Havana Canape Plate (Cock)—10" d.	7491	Briarton Covered Bowl—6-1/2" d.
7322	Plate	7492	Heatherton Covered Bowl
7324	Dorchester Tray—15" d.	7500	Rodney Desk Set Blotter Holder—12-1/2" x 20"
7325	Canape Plate (Savoy)—15" d.	7501	Rodney Memorandum Book—6-1/2" x 8-1/2"
7326	Bimini Canape Platter—15" d.	7502	Rodney Correspondence Rack—7 x 4"
7326	Regency Canape Plate—15" d.	7506	Rodney Correspondence Holder
7327	Canape Plate, Plain—15" d.	7506	Rodney Blotter Holder—19 x 24"
7328	Bermuda Plate, Plain—8" d.	7540	Dome Cigarette Box—4-3/4" h.
7329	Imperial Plate—8" d.	7541	Hexagon Cigarette Box—4-3/4" h.
7330	Malolo Plate—8" d.	7542	Penthouse Cigarette Box—6-1/2" l.
7331	Shell Canape Plate—10" d.	7546	Virginian (Buckingham) Cigarette Box—7-3/8" l.
7332	Hampton Oval Platter—10 x 15-1/4"	7547	Carolinian (Coach 'N Four) Cigarette Box--—11-3/4" l.
7380	Hyde Park Covered Vegetable Dish	7548	Manor Cigarette Box—7-3/8" l.
7380-B	Hyde Park Serving Dish	7549	Town House Cigarette Box—11-3/4" l.
7381	Dorchester Double Serving Dish—10" d.	7550	Piccadilly Cigarette Server
7382	Winchester Double Serving Dish—11-1/4" l.	7551	Hexagon Cigarette Box
7382-B	Manchester Serving Dish—11-1/4" l.	7604	Sir Walter Ash Receiver—5" d.
7383	Whitfield Double Serving Dish—9-1/4" d.	7604	Lucifer Matchbox Holder - 5 x 2-3/4"
7384	Canterbury Casserole—9-5/8" d.	7605	Crest Match Box Holder—1-1/2" x 2-1/2"
7400	Vanity Fair Candlestick—4" h.	7606	Vanity Fair Match Box Holder—1-3/8" x 1-5/8"
7401	Vanity Fair Candlestick—7" h.	7607	Hostess Ash Tray—2-5/8" d.
7402	Sherwood Candlestick—2-1/4" h.	7608	Hostess Set—4 each 7606 & 7607
7403	North Star Candle Holder	7609	Mall Ash Tray—5" d.
7404	Embassy Candle Holder	7610	Hexagon Ash Tray—3" d.
7405	Stratford Candle Holder—3-1/4" h.	7611	Prince Edward Candy or Nut Dish, Plain—5-3/8" d.
7407	Henley Flower and Fruit Bowl—10" d.	7611-A	Prince Edward Dished Ash Tray, Wreath—5-3/8" d.
7408	Berkeley Centerpiece Bowl—15" d.	7611-B	Prince Edward Dished Ash Tray, Medallion—5-3/8" d.
7409	Hemisphere Bowl—10" d.	7612-A	Royal Family Ash Tray, King—4-1/4" d.
7410	Wiltshire Bowl—9" d.	7612-B	Royal Family Ash Tray, Queen—4-1/4" d.
7411	Stratford Bowl—13-5/8" d.	7612-C	Royal Family Ash Tray, Jack—4-1/4" d.
7412	Laurel Bowl—16-1/4" d.	7612-D	Royal Family Ash Tray, Joker—4-1/4" d.
7413	Folkestone Bowl—7-3/8" d.	7613	Burleigh Ash Tray—5 1/2 d.
7414	Hampshire Bowl	7613	Ash Tray, Savoy—4-3/8" d.
7415	Sherwood Compote—10-1/4" d.	7614	Princess Anne Ash Tray, Plain—4-3/8" d.
7416	Corona Candy Dish—7" d.	7614-A	Princess Anne Dished Ash Tray , Wreath—4-3/8"
7417	Heather Candy and Nut Dish—3-1/2" d.	7614-B	Princess Anne Dished Ash Tray, Medallion—4-3/8" d.
7418	Thistle Candy Dish—4-3/4" d.	7614-P	Small Round Dishes—4-3/8" d.
7418	Cambridge Serving Dish—8-5/8" x 10-7/8"	7615	Tray
7419	Briar Bonbon Dish—6-1/2" d.	7615-A	Royal Family Snuffer, King
7420	Georgian Finger Bowl	7615-B	Royal Family Snuffer, Queen
7421	Croydon Jam Jar—6" h.	7615-C	Royal Family Snuffer, Jack
7422	Devonshire Candy Dish—5-1/2" d.		
7423	Cortland Jam Jar—5-1/2" h.		
7424	Lotus Bowl—5-1/8" d.		
7425	Cathay Bowl—5-1/8" d.		
7426	Ming Bowl—7" d.		

7615-D	Royal Family Snuffer, Joker
7616	Empire Smoker's Tray—6-1/4" d.
7617	Colby Ash Tray—4-3/4" d.
7620	Rainbow Matchmaster—6-1/8" l.
7620 1/2	Matchmaster Refills
7621	His Majesty's Cup—3" d.
7622	His Majesty's Porringer—5" d.
7623	Norwich Child's Cup—3" d.
7624	Norwich Porringer—5" d.
7625	Cape Cod Hurricane Lamp—7-3/8" h.
7691	Charleston Tobacco Jar
7692	Hatmaster Hat Brush—5-1/2" l.
7693	Clothmaster Cloth Brush—L6-1/2"
7694	Guards Military Brushes, Black Bristles---2-1/2" x 4-1/2"
7695	Guards Military Brushes, White Bristles---2-1/2" x 4-1/2"
7697	Raleigh Salt and Pepper Shakers—2-3/4" h.
7700	Snack Tray—9-3/4" d.
7701	Nibbler—8" d.
7702	Server—9-1/4" d.
7703	Bowl—10" d.
7725	Chip'n Dip Set—12-1/4" d.
7730	Serving Pitcher—2 qt.
G-9980	Altar Appointments—20" l.
10000-10331	Kensington Picture Frames

Kensington Moiré

Moiré is Silver; number with G is Gold

7740/G-7740	Norwich Canape—10" d.
7741/G-7741	Hastings Plate—10" d.
7742/G-7742	Canterbury Tray—10" d. 10-1/2" h.
7744/G-7744	Highgate Server—10" d. H4-1/4" h.
7748/G-1748	Stratford Server—8" d.
7749/G-1749	Stratford Server—9-1/4" d.
7750/G-7750	Carlton Server—8" d.
7751/G-7751	Carlton Server—9-1/4" d.
7752/G-1752	Georgian Tiered Tray—9-1/4" d. 10-1/2" h.
7753/G-7753	Norwood Compote—8" d.
7754/G-1754	Townsend Compote—9-1/4" d.
7758/G-7758	Savoy Tray—12-1/4" d.
7759/G-7759	Bradford Tray—12-1/4" d.
7760/G-7760	Glenwood Server—12-1/4" d.
7761/G-7761	Huntington Server—12-1/4" d.
7764/G-7764	Cumberland Square Tray—10-3/4"
7765/G-7765	Cumberland Tray with Handle

7775/G-7775	Winthrop Tray—10-1/4" x 16-3/8"
7776/G-7776	Winthrop Tray—11-5/8" x 18-7/8"
7777/G-7777	Winthrop Tray—13-1/2" x 21"
7780/G-7780	Westbrook Ash Tray—5" d.
7781/G-7781	Clinton Canape—15"

Kensington Ware Price List May 1, 1970:

7821/G-7821	Plymouth Tray
7822/G-7822	Bennington Tray
7823/G-7823	Concord Tray
7825	Wellesley Tray
7830	Trenton Tray
7831/G-7831	Yorktown Tray
7835	Brandywine Tray
7836	Camden Tray
7839	Princeton Tray
7840/G-7840	Saratoga Platter
7846/G-7846	Lexington Tray
7878/G-7878	Cornell Bread Tray
7883/G-7883	Monticello Tray
7890	Cornwall Chip & Dip Set
7891	Gallery Tray
7892	Yorktown Aspic Tray

Kensington Bent Glass

7915	Small Round Plate, Clipper Ship—6-1/2"
7916	Oval Celery, Band—6-1/4" x 13"
7917	Oblong Tray, Wheat—5-3/4" x 13"
7918	Round Canape, Eagle—10" d.
7919	Octagonal Plate, Stag—10"
7920	Rectangular Tray, Clipper Ship—7-7/8" x 14"
7921	Oval Tray, Flower—10-1/2" x 15-1/4"
7922	Zodiac Platter—15" d.
7923	Shell Platter—15" d.
7924	Shell Plate—11-1/2" d.
7925	Shell Plate—10" d.
7926	Shell Plate—8" d.
7928	Rectangular Tray, Stage Coach—14 x 24"
7929	Oval Tray, Tulips—13 x 23"
7933	Small Round Dish, Thistle—4-7/8"
7934	Small Round Dish, Rooster—5-3/8"
7935	Small Octagonal Dish, Arrow—5-3/8"
7946	Round Platter, Sunburst—21" d.
7950	Octagonal Plate, Flower—12"
7955	Eastern Hemisphere Plates—11-1/2" d.
7956	Western Hemisphere Plates—11-1/2" d.

A 1937 typewritten list without numbers or information and in no particular order (now alphabetized) includes the following Kensington objects:

Asparagus Server
Baby Book
Bedside Reading Lamp
Belt Buckle
Bottle Labels
Bracket Shelf
Candle Snuffer
Cheese or Cake Knife
Cheese Scoop
Cigarette Extinguishers
Cocktail Canape Server
Coffee Urn
Comb
Cookie Jar
Desk Thermometer
Dinner Gong
Dinner Bell
Double Candlestick
Electric Grill
Flower Pot Holder
Hand Mirror
Hot Dish Holder
Hot Food Warmer
Hurricane Candle Holder
Ice Guard
Jelly Spoon
Knitting Needle Case
Kleenex Holder
Lamps
Lemon Fork

Magnifying Glass
Muddler
Napkin Ring
Newspaper & Magazine Rack
Olive Pick
Olive or Pickle Fork
Orange Squeezer
Powder Jar
Paper Clip
Paper Cutter
Plate Warmer
Postage Stamp Box
Punch Bowl
Sardine Fork
Sconce
Scrap Basket
Silent Butler
Sugar Shaker
Swizzle Sticks
Tea Ball
Tea Strainer
Telephone Index
Thermo Jug
3-Part Vegetable Dish
Trophy Plaque
12" Ruler
Twine Holder
Window Sun Dial
Wine Cooler
Wine Glasses

Appendix B

Chase Brass & Copper Items by Lurelle Guild (from Kilbride)

200	Sparta Table Lamp
250	Hellenic Floor Lamp
350	Delphic Floor Lamp
500	Arcadia Table Lamp
501	Ulysses Table Lamp
502	Ionic Table Lamp
503	Flame Table Lamp
504	Luna Table Lamp
505	Plato Table Lamp
506	Doric Table Lamp
520	Diana Wall Lamp
521	Jupiter Wall Lamp
522	Athena Wall Lamp
523	Troy Wall Lamp
525	Thor Wall Lamp
526	Minerva Wall Lamp
529	Imperator Wall Lamp
530	Vesta Wall Lamp
531	Dolphin Wall Lamp
532	Olympia Wall Lamp
533	Sea Horse Wall Lamp
534	Thessaly Wall Lamp
540	Zeus Hanging Lamp
541	Paris Ceiling Lamp
543	Sybil Hanging Lamp
544	Apollo Ceiling Lamp

545	Aurora Hanging Lamp
546	Augusta Ceiling Lamp
601	Flying Cloud Table Lamp
640	Cyprus Ceiling Lamp
646	Planetarium Ceiling Lamp
27001	Individual Canape Plates (set of 4)
27002	Service Plate
27003	Sandwich Plate
27004	Cold Meat Platter
27005	Bread Tray
27007	Candlestick to 27012
27008	Circular Centerpiece to 27012
27009	Rectangular Centerpiece to 27012
27011	Electric Buffet Server (110 volts)
27012	Architex Adjustable Centerpiece
27013	Colonel Light
27014	Colonel's Lady Light
27026	Magazine Rack
27027	Newspaper Rack
27028	Fruit Basket
27029	Electric Buffet Server (220 volts)
28014	"High Hat" Jigger
28017	Jigger & Swizzle Set
90037	Niblick Swizzlers
90038	Pretzel Man
90046	Cups to Liqueur Set (Russel Wright Tray)
90047	Cups only to Liqueur Set
90088	Delphic Serving Spoon
90089	Delphic Serving Fork

Index